# A Bouquet of Dyson
and Other Reflections on Science and Scientists

# A Bouquet of Dyson
## and Other Reflections on Science and Scientists

Jeremy Bernstein
Stevens Institute of Technology, USA

**World Scientific**

NEW JERSEY · LONDON · SINGAPORE · BEIJING · SHANGHAI · HONG KONG · TAIPEI · CHENNAI · TOKYO

*Published by*

World Scientific Publishing Co. Pte. Ltd.
5 Toh Tuck Link, Singapore 596224
*USA office:* 27 Warren Street, Suite 401-402, Hackensack, NJ 07601
*UK office:* 57 Shelton Street, Covent Garden, London WC2H 9HE

**British Library Cataloguing-in-Publication Data**
A catalogue record for this book is available from the British Library.

**A BOUQUET OF DYSON**
**and Other Reflections on Science and Scientists**

Copyright © 2018 by World Scientific Publishing Co. Pte. Ltd.

*All rights reserved. This book, or parts thereof, may not be reproduced in any form or by any means, electronic or mechanical, including photocopying, recording or any information storage and retrieval system now known or to be invented, without written permission from the publisher.*

For photocopying of material in this volume, please pay a copying fee through the Copyright Clearance Center, Inc., 222 Rosewood Drive, Danvers, MA 01923, USA. In this case permission to photocopy is not required from the publisher.

ISBN 978-981-3231-92-4
ISBN 978-981-3238-28-2 (pbk)

For any available supplementary material, please visit
http://www.worldscientific.com/worldscibooks/10.1142/10762#t=suppl

Desk Editor: Christopher Teo

Typeset by Stallion Press
Email: enquiries@stallionpress.com

*I am grateful to Freeman Dyson for giving me
permission to publish these letters*

# Contents

| | | |
|---|---|---|
| **I** | **People** | **1** |
| 1 | A Bouquet of Dyson | 3 |
| 2 | The Pope | 17 |
| 3 | A Preprint | 33 |
| 4 | Murray | 41 |
| **II** | **Chronicles** | **49** |
| 5 | Houtermans | 51 |
| 6 | Charlotte | 61 |
| 7 | Einstein and the Fraud | 67 |
| 8 | Pontecorvo | 71 |
| **III** | **Science** | **75** |
| 9 | Three for the Road | 77 |
| 10 | Bode's Law and the Trappists | 85 |
| 11 | Advanced Quantum Mechanics | 89 |
| 12 | Gian Carlo | 103 |
| **IV** | **Nuclear Weapons** | **109** |
| 13 | An Error | 111 |
| 14 | Round and Round | 125 |
| 15 | Li6 | 127 |
| 16 | Is $E = mc^2$? | 131 |

| | | |
|---|---|---:|
| **V** | **Life** | **135** |
| 17 | A Trick of Memory | 137 |
| 18 | Checkers | 143 |
| 19 | A Little List | 147 |
| 20 | Anti-Semitism at Harvard | 157 |
| 21 | Twenty One | 161 |

My friendship with Freeman Dyson goes back over a half century. My first contact with him goes back to the late 1950s when I was at the Institute for Advanced Study and then evolved when I was a consultant at General Atomics in La Jolla, California. Freeman was then trying to design a space ship — the Orion — which would be propelled by atomic bombs. When I left the Institute Freeman and I continued our correspondence and I saved his letters. They are written in an almost calligraphically elegant handwriting. It is hard to see how you could make a mistake in a mathematical computation if you wrote that clearly. The letters show his human side and his enormous range of knowledge. I then pass on to two Italian scientists Enrico Fermi who is very well known and Bruno Pontecorvo who is not and my old boss Gian Carlo Wick. There are then two essays involving the physicist Fritz Houtermans who was an extraordinarily colorful character. There is a brief essay on Einstein's collaboration with a fraud. There is even an essay on the Titius Bode law and the new exo-planets. Because of my enduring interest in nuclear weapons the reader will find essays devoted to that.

# I People

# A Bouquet of Dyson

Freeman Dyson

The first time I met Freeman Dyson — at least in a manner of speaking — was in the fall of 1953. He had come to Harvard to give a series in the Morris Loeb and David M. Lee Lectures in Physics. Some of them were on a subject on which I was then doing my PhD thesis — the theory of mesons. My thesis advisor, the late Abraham Klein, suggested that we have a private talk with Dyson. I was very pleased with this idea because Dyson was already a hero of mine for his monumental work on quantum electrodynamics. His first lecture at Harvard had been about something else. He had been introduced by our local genius Julian Schwinger as "mister" Dyson which I took to be one of those professorial affectations. But I later learned that it was true. While he has been awarded honorary degrees from over twenty institutions Dyson never did bother to get his PhD.

He had been assigned a temporary office in the physics building which we had no trouble finding. The office had a couch on which Dyson was lying seemingly asleep. This did not bother Klein who began to drone on and on, oblivious to the fact that Dyson was lying there with his eyes closed. After some time, it registered on Klein that there was no response, so we left. I did not see Dyson for another four years by which time I had gone as a visiting member to the Institute for Advanced Study at which he was (and is) a professor. He occasionally had lunch with us at the Institute cafeteria. He rarely said much but seemed to be amused by the general patter. I never would have gotten to know him except for an accident. The accident consisted of the fact that we had taken the same late night train from New York to Princeton so we began to talk. As I recall, I had my automobile at the station, and I drove him home. He invited me in for a drink and wanted to know how, and why, I had gotten into physics.

*Toute proportion gardée*, our routes had been similar — mathematics. There was an important difference — he's a genius. It may have been on that occasion that I asked him what his earliest mathematical memories were. He said that when he was still young enough, for being "put down for naps", he began adding up $1 + 1/2 + 1/4 + 1/16\ldots$ and realized that the sum was approaching 2. In short he had grasped the idea of a convergent infinite series. At the time of our meeting he was separated from his wife. His children George and Esther were for the moment with her so he was living in this big house alone. We talked about that.

During the year, I saw him to talk, a few times. He was clearly very busy. One of the things he did was to read, and gloss over papers in

Russian journals which were not yet translated. Once when I went into his office he was relaxing in a canvas beach chair and reading the bible in Russian. That summer he went off to the General Atomics Company in La Jolla to consult, and I went to the Rand Corporation in Santa Monica to do likewise. It was not a success.

Although Rand had the superficial atmosphere of a college campus, it was devoted to the strategy of nuclear war. The Falstaffian figure of Herman Kahn was assuring everyone, that a few mega deaths in an exchange with the Soviets would be quite tolerable. I recall once going into a room with the other Rand physicists where a seismograph had been set up. We watched while it registered the quavers from a hydrogen bomb test in the Pacific. I found the whole atmosphere very depressing. In the meanwhile, the secretary we had in our building at the Institute was forwarding mail along with the gossip. I learned that Dyson was designing a space ship, that he had been to a bull fight, and previously had been bitten by a dog. I wrote a note to him saying that if any of these three things were true, he was having a better time than I was. Much to my surprise, a day or so later the phone rang, and it was Dyson inviting me to come down to La Jolla. I jumped at the chance.

It turned out that all three of these things were true, and the space ship, which was supposed to be powered by the exploding atomic bombs, was called the "Orion." It is not an accident that this was the name of a space ship in Kubrick's *2001*. There was more. He had been stopped by the police for walking. They had some reason. He had broken his glasses and was wearing scuba diving goggles to assist his vision. When asked for identification, he produced a card with his picture and finger prints on it from the Department of Defense. It said that the bearer of this card was entitled to receive top secret information. One can only wonder what went through the police officer's mind.

One of the considerations in designing the Orion was how much radiation would propagate through the superheated material following an explosion — the opacity. Dyson had proven a beautiful theorem that showed that quantum mechanics imposed a maximum. My job was to test this with numerical computations, which I did, on a now antique electro-mechanical calculator. Some fifty years later, we published this work. In the meanwhile, a great deal had happened in his life, much of which began that summer.

When I arrived in La Jolla, the children were still with their mother. She and Freeman were divorcing and the children were to come and live with him. When they arrived, they were accompanied by an attractive German woman named Imme Jung, who was to act as their governess, allowing Freeman the opportunity to go to work. He was totally preoccupied with the space ship, and wrote a stream of reports on everything, from the functioning of springs to how to shape the explosive force from an atomic bomb. As the summer evolved two complications developed. As Imme was scheduled to return to Germany, a second German governess named Margot had arrived as her replacement. But in the meanwhile, Freeman and Imme had fallen in love. Their happiness was evident to everyone. The four of us took to going out together, on one occasion driving to Baja California, and on another to visit some people I knew who had a ranch near Santa Barbara. By the end of the summer, Freeman and Imme had decided to get married. He was going to stay in La Jolla, and I was going to return to Princeton.

We began to write letters. I saved his, all written in an almost calligraphic script. The first one is dated November 25, 1958. And the last one I have is dated December 26, 1990. We then stopped communicating by letter and email took over. I have dozens of emails from Freeman. Some of them are very interesting but they lack the human touch of an actual letter. Can you imagine Lord Chesterfield's "Tweets to His Son?" Some of the letters are technical. One of them dated August 21,1981 begins "Sorry my friend you goofed." I had written a book review for the New Yorker, which had dealt with maps, and had carelessly described the Mercator projection — which projects the spherical earth onto the plane of a map — as "linear." Dyson writes, *"Mercator's is not a linear projection but a logarithmic one. So far as I know there is no geometric model for it. Precisely that was the originality of Mercator's idea."* He then goes on to provide a mathematical proof, and what is typical, he goes on to say, *"The really extraordinary thing about Mercator's projection is that it gives useful maps up to very high latitudes. This is possible only because the latitude scale is logarithmic."* He shows it works well, up to a latitude of 85°, and remarks that the *"map includes more than 99% of the globe. No geometric projection could possibly do as well as this,"* and then adds, *"it is a pity that schools do not teach geography anymore."*

He was of course right, and my error also slipped by the New Yorker's vaunted checkers. The interested reader can find the details on the web.

Many of the letters are anecdotal. Typical is the first one dated November 28, 1958. It was written just after Freeman and Imme had gotten married. It was addressed to "Jerry and Jane." Until I began writing for the New Yorker, I was always known as "Jerry" instead of my real name "Jeremy." "Jane" refers to Jane Kane, who was the secretary who had kept me informed about Freeman's activities while I was at Rand.

*"Dear Jerry and Jane*
*Your telegram made us very happy when it arrived on the wedding morning.*
*The wedding was a merry occasion. The judge took us in a break between two traffic violation cases and the children found this highly entertaining. We sat in the courtroom with the flowers and the 3 children* [**two of his own and one of his ex-wife's**] *and our witnesses...while the lawyers cross-examined some poor fellow who had been crashed into by a car-load of drunk Mexicans. Finally, the jury filed in, pronounced the verdict of "Guilty", and filed out, the judge heaved a sigh of relief and married us then and there.*

We flew to San Francisco for a 2-day honeymoon. This was only the hors d'-oeuvre. The real honeymoon shall take us to Mexico City in December. San Francisco was as beautiful as ever.

The most anxious moment of this week was 2 nights before the wedding when Imme and I went to a movie. Imme was trying on her ring and dropped it on the floor into a ventilator. Like the perfect gentleman I am, I crawled down outside the building and along about 100 feet of pitch-black and filthy and extremely narrow passages while Imme hammered on the floor above to guide me. I actually found the ring and got out alive, much to my surprise. I believe in the Middle Ages they had similar ordeals for young men who wanted to get married."

It is worth noting that in his Cambridge University days, Freeman had done some night climbing of buildings on the campus — an old tradition.

On February 6, 1959 I received a letter from Freeman that contained the following paragraph.

"Last Sunday the Mexican Federal Police raided the Club Deportivo Panamericana [**a place in Baja California we had visited**] collected 63 Americans, confiscated all their cash, and left them to languish in the Tijuana town jail until they could raise the $1600 bail each. The San Diego paper has been keeping this outrage on page 1 all the week. I whistle softly to myself when I think of Imme and Margot sharing a cell on one side of the corridor, you and I on the other. There, but for the grace of God.

By the way, Margot left us last week and found herself a job with a family in La Jolla who have 5 children under 10. So now she is finding out what real Americans are like."

Freeman and Imme had four daughters. On September 1970, Freeman wrote to me about an exchange he had, with his then seven-year-old daughter Miriam. The late Arthur Wightman was a professor of physics at Princeton, and was noted for his work on the mathematics of the quantum theory of fields. Freeman writes,

> "This morning I had an illuminating dialog with my 7-year-old daughter Miriam. At lunch yesterday Arthur Wightman scribbled some equations on a paper napkin and I took it home. So the following conversation occurred.
> Miriam. What is this?
> Me. It's a problem which Wightman gave me.
> Miriam. It is hard?
> Me. It was too hard for Wightman so probably it is too hard for me.
> Miriam. But Daddy, you know that negroes are just as smart as white people"

In 1986, I published an article in the New Yorker about climbing Popocatapetl in Mexico. On December 18 Freeman wrote,

We have been enjoying your "Breaking in at the New Yorker" and even more your Popocatapetl. We have vivid memories of our honeymoon in May 1959 when Imme, then 5 months pregnant, drove us in a dilapidated rental car up from Amecameca to the shoulder between Popo and Inxti. [**Iztaccihuatl is the slightly lower twin volcano.**] In those days the road was unpaved and there were no climbers. Not even any tourists. The sole human being up on the pass was a young Indian girl collecting bundles of tall grass reeds, presumably to be taken down to the village and woven into hats and baskets. Imme spoke to the woman in Spanish but she did not answer. We wondered how the hell she got up there, and how the hell *she would get down again. Probably by walking both ways. And we were feeling quite proud of our bravery getting up and down by car. Leaving her alone up there, when we drove down in the evening, we did not feel like heroes anymore.*

*What I liked about your story was the brevity of the climb itself. In the story, as in real life, it is the preparations that are memorable.*

*Happy New Year!*
*Imme and Freeman*

Freeman was away from the Institute during some of the spring of 1958. He was working on the space ship. So he missed the mystery. Threatening notices were being placed on the doors of the members. One might have dismissed this as a joke, except that Robert Oppenheimer, our director, had lost his clearance not too long before, and had been accused of being a Communist and possibly a spy. After Freeman returned I told him about what had happened. His first response was to accuse me. I said that this was not possible, because the hand writing on the notes was too good. Then he said, without explanation, "gaudy night." It was only a few years later that I understood the reference. In 1935, Dorothy L. Sayers published a Lord Peter Wimsey crime novel with that title. It deals with the kind of note leaving on Oxford, but of course there is murder involved. Once I saw the connection, I read the novel but was surprised by its casual anti-Semitism. A young man is asked if he is related to Lord Peter. "Why of course," said the young man sitting up on his heels. "He's my uncle; and a dashed more accommodating than the Jewish kind," he added, as though struck by a melancholy association of ideas. No explanation was offered for this bizarre association, which seems in your face (defiantly confrontational; blatantly aggressive or provocative). I wrote a note to Freeman from Oxford where I was spending a year, pointing this out. On February 3, 1972 he replied.

"Dear Jerry

Thanks for your note about Gaudy Night. I had totally forgotten that. I have been hoping some time to see 'Le Chagrin et la Pitié'" but it didn't come here yet. I am now reading "The Double-Cross System". It is the best of the WW2 spy books. The Israelis now have a superb opportunity to make use of its technique to deal the Soviet intelligence apparatus a mortal blow. I hope they are making the most of it."

Whatever it was, the intelligence apparatus seems to have survived. The mystery of the notes at the Institute was solved, when the police and possibly the F.B.I. began interviewing people. The young son of one of my physics colleagues confessed. He said it had been a "joke".

The next rather long letter requires some explanation. It is dated December 15, 1971. The first part refers to the fact that by 1971, I had made a few trekking trips to Nepal and had seen most of the great mountains there. In 1969, I had driven to Pakistan and had seen many of the mountains there. The rest of the letter refers to the fact that I was in the midst of writing a profile of Albert Einstein, which eventually appeared in the New Yorker. I had mentioned that I was doing this to Freeman.

Like everyone who writes about Einstein, I began my research with his "Autobiographical Notes" first published in 1949. In them, he tells us that at age sixteen, he had hit on a "paradox" which ten years later led him to the theory of relativity. He imagined that he could move as fast as a light wave. He could then attach himself to a crest or trough, and the wave would no longer appear wave-like. Hence he would have a way of determining his speed in an absolute sense. This he felt was impossible, so something had to be wrong. It turns out that relativity forbids motions of a massive object at the speed of light. Einstein does not put things quite this way, but that is the idea. What Dyson pointed out to me was that at age fifteen, Einstein actually wrote a little paper about this, and that in this little paper there was no hint of relativity at all. Here is the letter.

"Dear Jeremy

Thank you for letting us see the Einstein chapter which Helen gave us yesterday. [**This is a reference to Helen Dukas who had been Einstein's secretary in Germany and emigrated with him. She, along with Einstein's sister, shared his house in Princeton.**] I returned it to you, but I would be glad to have a copy to keep when you have one available. Also thanks for the Nanga Parbat picture which is superb. [**Nanga Parbat is a 26,600 feet tall mountain in Pakistan.**] I have three Nanga Parbat books, by Dyhrenfurth, Herrligkoffer and Buhl [**who first climbed the mountain**] and the whole thing has for me a mythical quality. The fact that you have looked down [**from a plane**] on that Olympian summit plateau makes you almost a mythological character too.

To return to Einstein. Recently I read the article "Über die Untersuchung des Aetherzustundes im magnetischen felde" [**"On the investigation of the aether state in a magnetic field"**] which Einstein wrote at the age of 15 and I wonder if you have read it. It appeared finally in "Physikalische Blätter" Vol. 27, 385 (1971). I found it extremely illuminating. In fact, it has very much the same value for the understanding of Einstein's development as the decipherment of Linear B had for the understanding of Greek history. In both cases, we were suddenly and unexpectedly confronted with a written document from a period far earlier than anyone believed possible. In both cases, the publication of the documents was attended with enormous complications, with Sir Arthur Evans and Otto Nathan playing possibly analogous roles.

My own reaction to Linear B and to the Einstein article went through the same three stages. First, with great enthusiasm and hope that some great mystery would be revealed.

Second, deep disappointment that the documents turned out to be so completely ordinary. Thirdly, understanding that precisely ordinariness of the documents <u>was</u> the mystery, making the miracles that came later even more miraculous.

When one looks in detail at the Einstein article, a number of things become immediately clear. His main concern was to find some experimental way of checking whether the mechanical modes of the aether were true. He was obviously impressed by the elegance of Hertz's experiments and the contrast between the elegance of the experiments and the clumsiness of the mechanical models. But he never explicitly questions the mechanical picture.

The most remarkable thing about the article is that it contains not the slightest hint of relativity. Not a hint of the famous question of what you see when you travel along with a light-wave. But is does contain two things that were decisive for the later history. (1) For Einstein the electromagnetic field came first, before he had been exposed to the whole imposing apparatus of Newtonian mechanics. So when he found out that Maxwell and Newton were in contradiction it was natural for him to hang on to Maxwell and discard Newton. The orthodox nineteenth-century education had of course precisely the opposite emphasis. (2) Einstein at 15 was not yet thinking in terms of a Gedanken experiment. He was discussing what he thought of as a real experiment. But the kind of experiment he proposed would have naturally led him to consider others which would be genuinely Gedanken in character. The style of his thinking is already going in the direction which would lead him to 1905.

I hope you will in your account of Einstein give some weight to the 1895 article. I think it is important for putting his work in perspective, just as Linear B has been for Greek history.

For the last two months I have been in the middle of great personal dramas concerning Einstein. How Einstein would laugh if he could see the learned men squabbling over his relics. I would laugh too except that the drama is a tragic one. Meanwhile I have been reading over again Einstein's great papers, the 1905 and the 1915 and finding them as fresh and wonderful as ever.

I hope you have a splendid time in Oxford. I probably won't see you before you go. But let's keep in touch.

Yours ever
Freeman"

The standard view in the 19th and early twentieth century, for the propagation of light, was that it propagated in a mechanical medium. Which was known as "ether", just as sound waves propagate in a medium. What Freeman was noting was that at age fifteen, Einstein accepted this and even proposed an experiment to detect the ether. In this paper there was no hint of what he would later write in is relativity paper *"The introduction of a "luminiferous ether" will prove to be superfluous…"* The "squabble" Freeman refers to had to do with the trustees of the Einstein estate blocking the publication of his letters. I had such an encounter with one of them, Otto Nathan. He held up the publication of my New Yorker profile of Einstein, by making all sorts of demands before he would allow us to use quotations. At one point, he told me that the estate had the rights to the formula $E=mc^2$. I finally had to pay a good deal for these rights. Soon after the first part of the profile had been published, Nathan called me to complain that he had not been credited.

Here is a letter dated October 14, 1968 just after I had taken a job at the Stevens Institute of Technology.

Dear Jerry
I was considerably moved by your letter. I had a very good feeling about Stevens when I went there for a day last winter, and I hoped you would see it the same way. It seems you did…I knew you were going through a rough time. Everything in your letter increases my respect for Stevens and for you.

*The last year was also a restless one for me... I did this year teaching at Yeshiva and I was seriously considering taking a permanent position there. I had come to one of those middle-age crises, feeling maybe I am too old to produce ideas anymore and I ought to settle down to a teaching job before it is too obvious.*

*Anyhow I finished the year at Yeshiva and came back here. And I have been in the full tide of happiness these last days, proving that one-dimensional Ising models with an interaction going like distance$^{-\alpha}$ have a phase transition[1] when $1 < \alpha < 2$. Now it is midnight and I just wrote the last words of the proof in a strangely luminous state of mind, as Gibbon describes in his autobiography how he wrote the last words of the "Decline and Fall."*

*So I am not too old after all, and still have something to do here. As you say, it is nice to know that given the choice there really is no choice.*

*I would much enjoy another visit to Stevens now you are there. The Nepal pictures are still to be seen, 2001 to be discussed, and much else. Your piece about nuclear weapons is good. It is difficult to write long sentence with clarity, but you did.*

*Love from us all*
*Freeman*

The Ising model is a mathematical model that describes how certain solids can become magnetic at critical temperatures. The simplest such model would be a string of tiny magnets which can point up or down. If at some critical temperature all the tiny magnets line up, then the string becomes magnetic. What Freeman showed was that, given the conditions of his proof, there must be such a critical temperature for a one dimensional string. It was a *tour de force* bit of mathematics.

---

[1] as conjectured by Kac

This final letter is dated May 6, 1969 and is written from the faculty club of the University of California at Santa Barbara.

*Dear Jeremy*

*This evening Shawn* [**William Shawn the editor of the New Yorker**] *telephoned and said he will print my stuff. Naturally I was pleased. I write now to thank you for your services as a midwife. Also thank you for the invitation to talk at Stevens without which this would never have happened.*

*I promised to send Arthur Clarke a copy of the lecture on space travel, but I don't have his address. If you happen to have it, please send it on a postcard.*

*By a strange coincidence, so soon after writing this piece for the New Yorker, I had again a close encounter with violent death. It came like the Hiroshima bomb on a peaceful sunny day. I was woken by a shattering explosion, followed by moans and cries for help. I was too scared and stunned to run down immediately. While I hesitated a man was burning to death. By the time I came down he had already dragged himself into the children's pool below my window and put out the flames. If I had come down sooner, I might well have been able to save his life. He died in hospital two days later.*

*The bomb was presumably intended only to burn down the Faculty Club where I am staying. Unfortunately, the caretaker found it and it blew up in his hands.*

*This campus is now very quiet, trying to digest what has happened. The blood and ashes around the pool have been washed away and the children are again splashing in it. And I am once again a survivor with a bad conscience.*

*All the best*
    *Freeman*

To this day the murder of the custodian Dover O. Sharp remains unsolved.

# The Pope 2

First published in Inference, Vol 2. Issue 4. December 2016
www.inference-review.com

Appeared as a book review of *The Pope of Physics: Enrico Fermi and the Birth of the Atomic Age*

by Gino Segrè and Bettina Hoerlin

Enrico Fermi came to Harvard to give the Loeb Lectures in the fall of 1953. I was eager to meet him. I admired his work, of course, but I also thought there might be a distant family connection between us. My aunt had given me the impression that after Fermi's arrival in the United States in 1939, she and members of the Fermi family had become the best of friends. When I ran into Fermi in the hallway of the Harvard physics building, I mentioned my aunt. Fermi gave me a chilly stare, and, without saying a word, walked away. Some years later, I described this encounter to someone who knew Fermi very well. He was not surprised.

During his visit, Fermi was persuaded to give an informal talk to a journal club formed under the guidance of Roy Glauber. Then a young assistant professor, Glauber would later win a Nobel Prize. He had gotten to know Fermi at Los Alamos during the war. I had hoped that Fermi would discuss the meson experiments being conducted at the University of Chicago. His talk went no further than an introduction to an elementary problem in the quantum theory. Most of us could have given the same lecture. With the exception of Paul Martin, we remained silent. Martin was the most brilliant of the graduate students; he objected to the approximations Fermi had made. Fermi gave a second lecture. Martin was still not satisfied. And a third. At that point, Martin gave up. Fermi would have continued until he had beaten Martin into submission.

*The Pope of Physics* is an account of Fermi's life and times. Gino Segrè and his wife, Bettina Hoerlin, have written their account from the inside out; they knew a good many people who knew Fermi. Hoerlin's father, Herman Hoerlin, worked with Fermi at Los Alamos, and Segrè's uncle, Emilio Segrè, had been one of Fermi's original collaborators in Rome. Both Segrè and Hoerlin could regard Fermi as a familiar presence.

Enrico Fermi was born in Rome on September 29, 1901. At the age of seventeen, after doing brilliantly in his entrance exams — sound waves, partial differential equations, Fourier transforms — Fermi was admitted to the *Scuola Normale* in Pisa.[1] Italian physics was not at the time illustrious. Fermi was largely self-taught; he knew more physics than his professors.

After graduation, Fermi came to the attention of the physicist Orso Corbino, who happened to be connected to the political powers controlling the universities.[2] Corbino recognized Fermi's exceptional abilities, and, until his death in 1937, remained Fermi's loyal *pistone*. Corbino arranged for Fermi to study at Göttingen. It was in Göttingen that David Hilbert presided over the world's most important center of mathematical research; but Göttingen was a jewel with two facets. The eminences of theoretical physics were either in Göttingen or passing through it: Max Born, Werner Heisenberg, James Franck, Pascual Jordan, and many others.[3]

Fermi never had any interest in pure mathematics, and did little to cultivate the mathematicians.

He did little to cultivate the physicists either. It is not, in fact, clear what Fermi was doing or whom he was cultivating.

Curiously enough, one of Fermi's early papers was purely mathematical.[4] Consider a sphere. A great circle on the sphere is known as a geodesic. It is the shortest distance between two points. In regions close to the geodesic, the sphere is locally flat. Curvature only becomes apparent globally. General relativity is a four-dimensional theory — three of space, one of time. Fermi considered an observer falling freely along a temporal geodesic; he is following the shortest distance between two points in time. It is possible to choose this geodesic so that it defines the spatial coordinates of the system at all points, and to choose it, moreover, so that its tangent vector defines the direction of time. Fermi gave an account of the local coordinates close to the temporal dimension, showing that they must be Euclidean. These coordinates still appear in relativity texts. There is no controversy about them.

The same cannot be said about one of Fermi's other papers. In 1905, Albert Einstein published a brief paper entitled, "Does the Inertia of a Body Depend Upon Its Energy Content?"[5] In it, he analyzed a situation in which a body emits an equal amount of radiation in opposite directions. In modern terms, we might think of these as light quanta, or photons. These quanta carry energy and momentum. Einstein argued that this energy is equivalent to the mass lost by the emitting object. The ensuing equation, $E = mc^2$, is now famous.

How general is this equation?

In 1904, the brilliant Austrian physicist Fritz Hasenöhrl claimed that under some circumstances, $E = 3/8 mc^2$. Still another physicist, Max Abraham, argued that $E = ¾ mc^2$ when the electron's self-interaction is taken into account.[6] Fermi was convinced that Einstein's formula was always correct. His paper was first published in Italian, translated into German, and reprinted in the *Zeitschrift für Physik*.[7] It was read by many physicists. The consensus now is that while Fermi may have been right in that Hasenöhrl was wrong, $E$ is not always equal to $mc^2$.[8]

Abraham, at least, had a point.

There was little or no culture in Fermi's family home, and, apart from what he had been taught in school, I doubt that he ever read a book of fiction or poetry for pleasure. It is interesting to compare him to Robert Oppenheimer, who read omnivorously in any number of languages, Sanskrit among them. Fermi never liked him very much. I was a postdoctoral fellow at the Institute for Advanced Study in Princeton between 1957 and 1959. Oppenheimer was then the director. From time to time, junior fellows would be summoned to his office and quizzed on what they had been doing. It was better to say you had been doing nothing than to tell Oppenheimer something incorrect or trivial. When I got my summons, I was not doing physics: I was reading Marcel Proust. I told Oppenheimer the truth. He gave me a wistful look. When he had been my age, he said, he had taken a bicycle trip around Corsica. At night he read Proust by flashlight. I found this endearing. Fermi would have found it irritating.

His time in Göttingen at an end, Fermi had no job. Corbino came to his rescue. He arranged for Fermi to teach a course at the University of Rome. By living at home, Fermi got by. He then accepted a lectureship in Florence, and, with Corbino's help, in 1927, a professorship in Rome.[9] That year, he married Laura Capon, the daughter of an assimilated Jewish naval officer.[10] The physics department was located on the Via Panisperna. The Boys of Panisperna, as they became known, changed twentieth-century physics.[11] Fermi was designated the Pope, and Corbino *Padreterno*, God Almighty, in recognition of his role in sustaining the group. Segrè's uncle Emilio was *Basilico*, the legendary basilisk that could cause death with a single glance. Just who served as Beelzebub is not known.

Ettore Majorana was known as *Il Gran Inquisitor*, because of his brilliant and corrosive questions.[12] He was the one member of the group that Fermi considered his intellectual equal. Fermi arranged for Majorana to spend some time in Germany with Heisenberg. The visit unhinged Majorana. He wrote an anti-Semitic letter to Segrè, and when he returned to Italy, entered into a period of self-isolation that lasted until 1937, when Fermi used his influence to get him a professorship at the University of Naples. In 1938, Majorana disappeared while taking a ferry from Palermo to Naples. His body was never found.[13]

The neutron afforded Fermi his first great triumph. James Chadwick correctly identified the neutron in 1932. It had been missed, as Majorana pointed out, by Irène Curie and Frédéric Joliot, who thought they had seen a very energetic massless radiation quantum. Majorana was sure the particle had to be massive.[14] He was right. The work Fermi did on the neutron was both experimental and theoretical. The neutron is an unstable particle with a lifetime of a little less than fifteen minutes. Fermi constructed the first correct theory of its decay. Two of the particles the neutron decays into are very familiar: the proton and the electron.[15] The third, casually suggested by Wolfgang Pauli, was a very light electrically-neutral particle. The Italians call the neutron the *neutrone*, so Fermi named Pauli's particle the *neutrino*. The name stuck. Fermi's theory has evolved over time, but one of the things that has remained is the insight that the electron and neutrino do not exist within the neutron before its decay. They are created by the process.[16]

Soon after the neutron was discovered, physicists realized that it was an ideal nuclear probe. Since it was electrically neutral, it could penetrate the nucleus of an atom without being repelled. Some nuclear collisions produce neutrons and these could be used as sources. Fermi and the Boys bombarded various elements with neutrons. They produced new, never-before-seen versions of matter, but they also made an accidental discovery that changed physics. They noticed that if their target element was on a wooden rather than a marble table, interactions with the neutrons were enhanced. At first Fermi thought this must be some kind of mistake, but then, one morning, he put paraffin in front of the target and got the same effect. He then went home for lunch and a siesta. By the time he returned to the laboratory, he understood everything.[17]

To help one appreciate Fermi's insight, I need to say something about the quantum theory nature of the neutron. The neutron is both a wave and a particle. So is a baseball, but because of its mass and size this fact is unobservable. If someone gives you the task of breaking a window with a baseball, you increase your chances by throwing the ball harder. Precisely the opposite is true of the neutron. Its size, as measured by its wave length, is increased if it is slowed down, and as the wavelength of the neutron becomes comparable to the size of the nucleus, its interaction with a nucleus becomes more likely. The effect of the paraffin, or the wood in the table, was to slow the neutrons down. To moderate the speed of a neutron, one wants a collision with a particle of comparable mass. Paraffin is rich in hydrogen, which contains protons of mass comparable to the neutron. So paraffin acts as a moderator of neutron speeds. Moderators play an essential role in the design of nuclear reactors.

The Boys studied one element after another, coming eventually to uranium. Fermi had a clear idea of what he was going to find. He was sure the neutron would enter the uranium nucleus and produce an unstable uranium isotope, which would transform itself into a nucleus with an additional proton. This would be a new transuranic element. It would be produced in micrograms, but its radioactivity would give it away. When Fermi and the Boys did the experiment, they found novel radioactivity, and were sure it was from the transuranic element. Fermi's result was published in *Nature* in 1934.[18] Not long thereafter, a German chemist named Ida Noddack published a paper in a chemistry journal arguing that Fermi's experiment had not considered all the possibilities. The nucleus might have broken up into several large fragments. In modern terminology, it might have fissioned. This paper became known to Fermi. He chose to ignore it.[19] Was it because Noddack was a woman? I do not think that this was an important reason. If someone like Lise Meitner had made such a claim, it would have been taken very seriously. But Noddack did not give any mechanism for the fission. She did not even ask whether energy conservation allowed this to happen. When Meitner and her nephew Otto Frisch considered similar experiments done by Otto Hahn and Fritz Strassmann a few years later, the first thing they did was to study the energy balance.[20]

I once asked Emilo Segrè why the Boys did not discover fission. It was an accident, Segrè explained. Radioactivity from the bombarded uranium was overwhelming their detection equipment, so they introduced extra shielding around the uranium. When a nucleus like uranium fissions, the resulting charged energetic nuclear fragments leave an impact on any detector. But in this case, they were blocked.

There is a remarkable twist to the story. Fermi won the Nobel Prize in 1938. A note added to Fermi's Nobel lecture reads:

The discovery by Hahn and Strassmann of barium among the disintegration products of bombarded uranium, as a consequence of a process in which uranium splits into two approximately equal parts, makes it necessary to reexamine all the problems of the transuranic elements, as many of them might be found to be products of a splitting of uranium.[21]

Noddack had been right after all.

Fermi was non-political, almost to the extreme. One has the impression he did not much care who ran the government so long as he was free to do his physics. Fermi joined the Royal Italian Academy, a creation of Mussolini which required members to wear an elaborate uniform at meetings.[22] But it helped finance the physics that was all that mattered to Fermi.

I have the impression, from this book and elsewhere, that if his wife had not been Jewish, Fermi would have done what Gian Carlo Wick and others did and remained in Italy, hoping Mussolini would lose the war. Following Hitler, Mussolini introduced racial laws into Italy in the late 1930s. Fermi could see great Italian mathematicians being thrown out of universities;[23] he began to wonder how safe his family was. He began a discreet search among American universities for a job. One was offered to him at Columbia. He asked them, as a matter of safety for his family, not to announce that this was a permanent job. He had his children baptized, so there would be no problems with their passports. Without being told anything, Niels Bohr guessed what Fermi was going to do. He told Fermi in confidence that he was going to win the 1938 Nobel Prize, realizing that this might provide an exit strategy along with some useful money. Fermi won $45,000, which was indeed very useful. The family had no problems getting to the United States and starting a new life.[24] But fission changed everything.

Many physicists made the elementary calculation of how much energy would be liberated if one fissioned a kilogram of uranium. Most of them regarded the calculated amount as science fiction. One person who took it all seriously was the Hungarian-born polymath, Leo Szilard. Szilard had attached himself like a buzzing insect to the Columbia physics department. What he was buzzing about was the possibility of nuclear weapons.

In the early 1930s, Szilard invented the notion of a chain reaction: if a reaction produced a small amount of energy and a method of reproducing itself, this energy could be amplified. He even took out a patent. When fission was discovered, he immediately saw how his idea could be realized. Fission is neutron induced. If, in addition to the fission fragments, two or more neutrons are produced, these can in turn induce new fissions, and on and on. If this process were slow, it might be an interesting laboratory phenomenon, but it would not constitute an explosion.

How fast was it?

Uranium fission produces nearly three neutrons, on average. These travel at about a tenth of the speed of light, meaning that a whole kilogram could be fissioned in about a microsecond. This is indeed an explosion. The realization drove Szilard into high gear.

Szilard was desperately afraid the Germans might make a bomb before the U.S. He managed to persuade the Belgians not to sell the Germans uranium from the Congo. He also managed get American physicists to agree not to publish anything about fission.[25]

Fermi realized that in order to build a bomb, physicists would first have to create a device that could realize a self-sustaining chain reaction — a reactor, to use the current term. After some discussion, it was decided that this would be done at the University of Chicago.[26] By June 1942, Fermi and his family had moved to Chicago.

Segrè and Hoerlin give an excellent account of how the first reactor was built and successfully tested on December 1, 1942.[27] That this took place was largely due to Fermi's genius for practical engineering physics. I would like to amplify one thing, which illustrates why the German project failed. The reactor needed a moderator. Graphite was chosen because it was plentiful and easy to work with. Szilard, who had joined the project, learned how graphite was produced by the National Carbon Company. He discovered that they added boron to help with structural stability. As Szilard knew, boron was a neutron absorber; all their graphite had to be boron-free. The Germans also wanted to use graphite as a moderator. One of their few remaining nuclear experimenters, Walther Bothe, took on the task of seeing how graphite reacted to neutrons. It never occurred to him to ask about boron. He simply concluded that graphite would not work. The German researchers then tried to use heavy water, but could never get enough. It was one reason why the program failed. Afterwards, they accused Bothe of having made a mistake. He did not make a mistake; his result was correct, given the graphite he used. No one felt they could question the result. The mistake was in discouraging the free flow of ideas.

Los Alamos began functioning in the spring of 1943 and the Fermis went there in August of 1944. Laura Fermi wrote a charming book about the experience, *Atoms in the Family*, parts of which were published in *The New Yorker*. Life in Los Alamos was difficult for the families; they had no idea why they were there. It probably helped create a rift between Fermi and his son Giulio, which never really healed.

The story of what Fermi did at the first test at Trinity in New Mexico is legendary.[28] The assembled physicists tried to guess the bomb's explosive power. To understand what Fermi did, one must understand the time sequence. I myself witnessed two tests in the summer of 1957. First, there is the light, brighter than a thousand suns. Then comes the supersonic shock wave, which makes your ears hurt. Then finally comes the noise. Fermi realized that if he could measure the strength of the shock wave he would have a measure of the strength of the explosion. He tore strips of paper and, when the shock wave arrived, let them drop. By seeing how far they went he was able to make a pretty accurate estimate of the strength of the explosion.

When the war ended, Fermi gave a mini-course on nuclear physics, which later morphed into a famous course he gave at the University of Chicago.

He also gave a classified lecture on the physics of Edward Teller's then-version of the hydrogen bomb: the classical Super. There is some irony in this, since when Fermi first suggested to Teller, well before Los Alamos, that a nuclear fusion bomb might be possible, Teller thought the idea was crazy. He later changed his mind and became obsessed.[29]

Fermi's lecture was a careful analysis of the physics, concluding with his feeling that there were still a great many uncertainties. One member of the audience was Klaus Fuchs, both a member of the British delegation and a Russian spy. Fuchs managed to turn over a copy of this lecture to the Russians. I think the main thing it did was to inform them of the status of the U.S. program. For many years, the lecture was classified but eventually the Russians released it and you can find it on the web.

After the war, Fermi was necessarily thrown into the political arena when it came to nuclear energy. He was a member of the General Advisory Committee of the Atomic Energy Commission. In 1949, they were asked to advise on a crash program to build the hydrogen bomb. Fermi and Isidor Isaac Rabi wrote an addendum to the report that read:

The fact that no limits exist to the destructiveness of the weapon makes its very existence and the knowledge of its construction a danger to humanity as a whole. It is necessarily an evil thing considered in any light. For these reasons we believe it important for the president of the United States to tell the American people and the world that we think it is wrong on fundamental ethical principles to initiate the development of such a weapon.[30]

This may well be the strongest political statement Fermi ever signed. They were right. The hydrogen bomb should never have been built.

Fermi now threw himself back into the new physics of high energy accelerators. One was built at the University of Chicago and was used to study the interactions of mesons and neutrons and protons.

Segrè and Hoerlin tell the story of Freeman Dyson's encounter with Fermi.[31] Dyson was then a professor at Cornell University, and had to provide dissertation problems for his students there. He decided to use a then-fashionable approximation method to try to reproduce Fermi's results. I am quite familiar with the method, since I used it in my own thesis. Its virtue is that it is fairly easy to calculate with and seems to give sensible answers. Its flaw is that the terms you leave out may be as big or bigger than the ones you include. In any event, Dyson and his students got results that seemed to agree with Fermi's experiment. Dyson then went to Chicago to see Fermi. It was not a long meeting. Fermi objected to the method and then asked how many free parameters Dyson had used to make the fit. Dyson said four. Fermi told him that John von Neumann had often said "with four parameters I can fit an elephant and with five I can make him wiggle his trunk." That was the end of that.

There is a story they do not tell, which has a much happier conclusion. By the early 1950s, a variety of new particles had appeared, first in cosmic rays, and then in accelerators. No one had predicted them and they were very confusing. To help characterize them, Murray Gell-Mann introduced a new kind of charge which he called strangeness. Familiar particles like the neutron and the proton had zero strangeness, but others could have plus or minus one or two. In the production of particles, strangeness had to be conserved, but in their decay it could be violated. There was a set of mesons called kaons. They came in a charge plus and minus and in a neutral particle and an anti-particle. The $K^+$ was assigned strangeness $+1$ while the $K^-$ was assigned the strangeness $-1$. The $K^0$ had the same strangeness as the $K^+$ and the anti-$K^0$ had the same strangeness as the $K^-$. It was a nice scheme and guided Gell-Mann's later construction of particles using quarks.

Fermi asked whether a $K^0$ might transform itself into an anti-$K^0$ by the weak interaction and, if so, what the consequences would be. My own guess is that Fermi knew the answer but gave Gell-Mann the pleasure of working it out himself, which he did, together with Abraham Pais. What happens is that the two particles oscillate back and forth in time, which can be observed by studying their decays. The paper contains the acknowledgement "One of us, (M. G.-M.), wishes to thank Professor E. Fermi for a stimulating discussion."[32] A bit of an understatement.

By the early 1950s, it was clear to people who knew him that there was something wrong physically with Fermi. He was first assured by a doctor that he was alright, but by the fall of 1954, it was obvious that he had an inoperable cancer. Fermi was totally serene about his impending death. People who went to see him came away shaken.[33] They could not believe that he was going to die. He did so on November 28.

## Endnotes

1. Emilio Segrè, *Enrico Fermi: Physicist* (Chicago: University of Chicago Press, 1970), 11–13; Gino Segrè and Bettina Hoerlin, *The Pope of Physics: Enrico Fermi and the Birth of the Atomic Age* (New York: Henry Holt and Co., 2016), 19–20.
2. Gino Segrè and Bettina Hoerlin, *The Pope of Physics: Enrico Fermi and the Birth of the Atomic Age* (New York: Henry Holt and Co., 2016), 29–30.
3. Gino Segrè and Bettina Hoerlin, *The Pope of Physics: Enrico Fermi and the Birth of the Atomic Age* (New York: Henry Holt and Co., 2016), 30–32.

4. Enrico Fermi, "*Sopra i fenomeni che avvengono in vicinanza di una linea oraria*" (On Phenomena Occuring Close to a World Line), *Rend. Lincei* 31, no. 1 (1922): 21–23, 51–52, 101–103. An English translation is also available.
5. Albert Einstein, "*Ist die Trägheit eines Körpers von seinem Energiegehalt abhängig?*" (Does the Inertia of a Body Depend Upon Its Energy Content?) *Annalen der Physik* 18 (1905): 639. For an English translation see Albert Einstein et al., *The Principle of Relativity*, trans. George Barker Jeffery and Wilfrid Perrett (London: Methuen and Company Ltd., 1923).
6. Max Abraham, "*Prinzipien der Dynamik des Elektrons*," *Annalen der Physik* 315, no. 1 (1903): 105–79 (1903). See also Stephen Boughn and Tony Rothman, "Hasenöhrl and the Equivalence of Mass and Energy," (2011), arXiv: 1108.2250v4.
7. Fritz Hasenöhrl, "*Zur Theorie der Strahlung in bewegten Körpern*," *Annalen der Physik* 15 (1904): 344–76. See also Stephen Boughn and Tony Rothman, "Hasenöhrl and the Equivalence of Mass and Energy," (2011), arXiv: 1108.2250v4.
8. See Stephen Boughn, "Fritz Hasenohrl and $E = mc^2$" (2013), arXiv: 1303.7162.
9. Gino Segrè and Bettina Hoerlin, *The Pope of Physics: Enrico Fermi and the Birth of the Atomic Age* (New York: Henry Holt and Co., 2016), 47.
10. Gino Segrè and Bettina Hoerlin, *The Pope of Physics: Enrico Fermi and the Birth of the Atomic Age* (New York: Henry Holt and Co., 2016), 56–62; also see her memoir, Laura Fermi, *Atoms in the Family: My Life with Enrico Fermi* (Chicago: University of Chicago Press, 1954).
11. Gino Segrè and Bettina Hoerlin, *The Pope of Physics: Enrico Fermi and the Birth of the Atomic Age* (New York: Henry Holt and Co., 2016), 65–69
12. Gino Segrè and Bettina Hoerlin, *The Pope of Physics: Enrico Fermi and the Birth of the Atomic Age* (New York: Henry Holt and Co., 2016), 69.
13. Gino Segrè and Bettina Hoerlin, *The Pope of Physics: Enrico Fermi and the Birth of the Atomic Age* (New York: Henry Holt and Co., 2016), 116–117; also see Josh Gelernter, "Lost at Sea," *Inference: International Review of Science* 2, no. 3 (2016).
14. Gino Segrè and Bettina Hoerlin, *The Pope of Physics: Enrico Fermi and the Birth of the Atomic Age* (New York: Henry Holt and Co., 2016), 84–86.

15. Strictly speaking it is an anti-neutrino that is emitted in the neutron decay. The anti-neutrino may or may not be the same as the neutrino. If it is, it is called a Majorana neutrino, since he was the first one to suggest this possibility.
16. Gino Segrè and Bettina Hoerlin, *The Pope of Physics: Enrico Fermi and the Birth of the Atomic Age* (New York: Henry Holt and Co., 2016), 89–90.
17. Gino Segrè and Bettina Hoerlin, *The Pope of Physics: Enrico Fermi and the Birth of the Atomic Age* (New York: Henry Holt and Co., 2016), 102–104.
18. Gino Segrè and Bettina Hoerlin, *The Pope of Physics: Enrico Fermi and the Birth of the Atomic Age* (New York: Henry Holt and Co., 2016), 105–107.
19. Gino Segrè and Bettina Hoerlin, *The Pope of Physics: Enrico Fermi and the Birth of the Atomic Age* (New York: Henry Holt and Co., 2016), 106.
20. Gino Segrè and Bettina Hoerlin, *The Pope of Physics: Enrico Fermi and the Birth of the Atomic Age* (New York: Henry Holt and Co., 2016), 127–31.
21. Enrico Fermi, "Artificial Radioactivity Produced by Neutron Bombardment: Nobel Lecture, December 12, 1938," in *Nobel Lectures, Physics 1922–1941* (New York: Elsevier Publishing Co., 1965), 417.
22. Gino Segrè and Bettina Hoerlin, *The Pope of Physics: Enrico Fermi and the Birth of the Atomic Age* (New York: Henry Holt and Co., 2016), 72–73.
23. For a list of twelve mathematicians who were removed from their positions in 1938, see Angelo Guerraggio and Pietro Nastasi, *Italian Mathematics Between the Two World Wars* (Berlin: Springer Science & Business Media, 2006), 262.
24. Gino Segrè and Bettina Hoerlin, *The Pope of Physics: Enrico Fermi and the Birth of the Atomic Age* (New York: Henry Holt and Co., 2016), 118–24.
25. Gino Segrè and Bettina Hoerlin, *The Pope of Physics: Enrico Fermi and the Birth of the Atomic Age* (New York: Henry Holt and Co., 2016), 147–50.
26. Gino Segrè and Bettina Hoerlin, *The Pope of Physics: Enrico Fermi and the Birth of the Atomic Age* (New York: Henry Holt and Co., 2016), 174–75.
27. Gino Segrè and Bettina Hoerlin, *The Pope of Physics: Enrico Fermi and the Birth of the Atomic Age* (New York: Henry Holt and Co., 2016), 184–98.
28. Gino Segrè and Bettina Hoerlin, *The Pope of Physics: Enrico Fermi and the Birth of the Atomic Age* (New York: Henry Holt and Co., 2016), 240–42.

29. Gino Segrè and Bettina Hoerlin, *The Pope of Physics: Enrico Fermi and the Birth of the Atomic Age* (New York: Henry Holt and Co., 2016), 274.
30. *The Pope of Physics: Enrico Fermi and the Birth of the Atomic Age* (New York: Henry Holt and Co., 2016), 279.
31. *Gino Segrè and Bettina Hoerlin, The Pope of Physics: Enrico Fermi and the Birth of the Atomic Age (New York: Henry Holt and Co., 2016), 273.*
32. *Murray Gell-Mann and Abraham Pais, "Behavior of Neutral Particles under Charge Conjugation," Physical Review 97, no. 5 (1955): 1,389.*
33. *Gino Segrè and Bettina Hoerlin, The Pope of Physics: Enrico Fermi and the Birth of the Atomic Age*

# A Preprint     3

"Reproduced from The American Journal of Physics, 85, 155, 2017 with the permission of the American Association of Physics Teachers."

I was recently straightening out a drawer in a cabinet, when I came across a somewhat decayed looking preprint of a paper entitled "The effect of the Hyperfine Splitting of a µ-Mesic Atom on Its Lifetime." There is no date given but I happen to know, as I was one of its authors, that it was in the spring of 1958.[1] My co-authors were T. D. Lee, C. N. Yang and Henry Primakoff. This remarkable grouping, which I am going to explain, is an illustration of the Russian saying that to live a life is not like crossing a field.

The events that led up to it began a year or so earlier in Cambridge, Massachusetts. I was then the "house theorist" for the Harvard Cyclotron, a job which allowed me to work on anything I liked. My interest was in the more phenomenological aspects of what was then the theory of elementary particles. What I would call, broadly speaking, the field theory community in Cambridge was then quite small. At Harvard, besides myself, there were Ken Johnson, Roy Glauber and of course Julian Schwinger. At M.I.T. there was Herman Feshbach and Viki Weisskopf, and Francis Low was then a visitor. We met for lunch more or less once a week on Wednesdays at Chez Dreyfus where Schwinger would present his latest ideas, usually on paper napkins, while the rest of us sat transfixed.

An exception was Francis who on one occasion, turned to me and said in a loud voice, "He is wrong. The man is wrong." He then explained to Schwinger.

Through these lunches I had gotten to know Weisskopf. After one of the lunches he explained that he had a task for me. The young Austrian physicist Walter Thirring was visiting M.I.T. and he was trying to write a new English version of his book on quantum electrodynamics. Viki felt he might need help with his English and said I was the person to do this. Viki also told me that he thought the German edition was a masterpiece, comparable to Pauli's little book on relativity written at about the same age that Thirring was then.

Viki was very persuasive so I met Walter and agreed to help. It turned out that Walter's written English was as good as mine, but we became friends and he was kind enough to give me an acknowledgement in his book. He also gave me a pre-print to read, that involved the hyperfine structure in μ-mesic atoms. I have since forgotten the conclusion of this preprint, but I was struck by the fact that compared to the hyperfine interactions in ordinary atoms, those of μ-mesic atoms were a couple of orders of magnitude larger. Basically, these larger interactions are due to the muon in the hydrogen atom ground state being closer to the proton by a factor of $m_\mu/m_e \sim 207$. I then forgot all about it when I began my first year at the Institute for Advanced Study in the fall of 1957. But I did bring Thirring's preprint along.

That year was an exciting one at the Institute. Lee and Yang won the Nobel Prize for their work on parity non-conservation. Lee had taken a leave of absence from Columbia and he and Yang continued their collaboration with a high volume of Chinese chatter, emanating from one or another of their offices. I devoted much of my time trying to learn about the weak interactions, especially by reading their papers. In particular, I looked at what they had written about muon capture by protons. I also happened to look at Thirring's paper. I noticed that Lee-Yang paper did not mention the hyperfine interaction and I wondered why. I had no idea what, if any, role it would play and even no idea of what to look into. Here it would probably have remained except for one of those coincidences — acts of chance that life sometimes deals you.

It was a lovely spring Saturday morning and I was in my office reading both a paper of Lee and Yang and the one by Thirring. I was due to spend a long weekend in New York and was going to drive in early in the afternoon. I spotted Lee walking across the lawn to his office. I barely knew him if at all but I thought that it might be a chance to ask him about the role, if any, of the hyperfine splitting in these muon processes. I intercepted him before he reached his office and asked if he and Yang had looked into this. Lee said that there were no consequences. I asked him if he would mind explaining that to me and we went to his office. He went to the blackboard and began computing at a furious rate saying things like "spin flipper." I had no idea of what he meant and I went to New York before I found out. When I came back on Monday afternoon I looked at my post box. I noticed what seemed to be a typed paper. It had three authors — Lee, Yang and me! It had the same title as the preprint I saved. I was completely astounded. I had done nothing but ask a question which I had no idea how to answer and somehow this paper had appeared. The speed at which they worked was incomprehensible to me and their generosity was overwhelming. I was a more or less unknown junior physicist and they would have been perfectly within their rights simply to thank me for asking a question. What they did made an enormous difference to me. I am sure it played a role in my getting a second year at the Institute and eventually a National Science Foundation Fellowship, which allowed me to spend two years in Paris. To live a life…

Before I discuss the contents of the paper, I must explain how Primakoff got his name on it. This was also a matter of chance. That very week there was a meeting of the American Physical Society in New York and I went back to the city clutching our paper. I have no idea who I was planning to show it to, but as luck would have it, I showed it to the one person who understood it without even bothering to read it. This was Valentine Telegdi, a brilliant and sardonic professor at the University of Chicago whom I had met on his visits to Princeton. He was known as "Mister muon" because of the experiments he had done. Telegdi told me that he knew all about this because of his contacts with Henry Primakoff, a deep thinking theorist then at Washington University. Telegdi said Primakoff had done the same work. Upon my return I told my two collaborators who were very surprised since nothing had been published. They called Primakoff and indeed he had considered the same problem and was preparing a manuscript which he would send. When it came, it was gigantic and festooned with the sort of baroque symbols Primakoff liked. I am not sure that he had suggested an actual experiment, which is what Lee and Yang did but he certainly had the same idea so he became a co-author. Later we became good friends. Although he had been born in Odessa he had a slight southern accent. Telegdi began referring to our paper as BLYP — pronounced "blip." Let me now explain our paper.

The foundational idea is the role of the large muon mass compared with that of the electron. This enhances all the hyperfine effects. The hyperfine interaction is the magnetic coupling of the magnetic dipole of the nucleus with the magnetic field generated by the spin and orbital angular momentum of the electron, or in this case the muon. For the muon, the splitting of the hyperfine frequencies is much larger than the inverse life time of the mu-mesic atoms. This means that the hyperfine states de-cohere and the capture rates from the two states are in general different, for reasons I will explain.[2] The muon lifetime is sensibly the same in the two states. There is a small wave function effect. What Lee and Yang had computed previously was the lifetime for which the hyperfine effect is irrelevant.

To compute the hyperfine effect, Lee and Yang had to invoke a nuclear model. In this model the muon is assumed to be in the k-orbit which is closest to the nucleus. The nucleus is supposed to have a non vanishing spin I. The two hyperfine states have angular momenta $I+\frac{1}{2}$

and I$-\frac{1}{2}$. For a single proton this would be the triplet and singlet *s* states. For simplicity our nucleus is assumed to consist of a single proton outside a core whose total spin is zero, so this proton carries the angular momentum L and the two hyperfine states have angular momentum L$+\frac{1}{2}$ and L$-\frac{1}{2}$. The capture by the core is the same for the two hyperfine states. Again for simplicity, the outside proton is taken to be free. If the capture interaction were spin independent, there would be no difference in the two rates. At the time our paper was written, the nature of the weak interaction was still being debated. So along with the vector and axial vector of today, we also considered and scalar and tensor coupling which makes the results more complicated than they need be. That Lee and Yang could sort all this out in a little over a day borders on the miraculous. Apart from the scalar, the other possibilities are all spin-dependent so they imply different capture rates for the two hyperfine states. This suggests an experiment, what turns out to be a very difficult experiment, the results of which were published by Roland Winston, a student of Telegdi's in 1963.[3] Because of the different capture rates of the muons in the two hyperfine states, the number of muons as a function of time in each state is different. Hence the decay electrons will not follow a simple exponential in time. Measuring the departure from this exponential was what was proposed and what Winston measured. As far as I can tell, by a study of Google, other methods of measuring the hyperfine interaction have since been carried out and much more sophisticated nuclear models have been used. Primakoff published his gigantic paper as a review article.[4] I got to know him and even work with him a bit. He ended his career at the University of Pennsylvania, He died in 1983, Walter Thirring in 2014.The other three authors of the BLYP are still functioning. I might mention that, that spring, Bohr happened to be visiting the Institute. Oppenheimer arranged a little seminar where the various members were asked to speak about their work, I was asked to talk about what I had done with Lee and Yang. They had much more important things to tell Bohr. I was very nervous and spoke very fast. Bohr listened impassively and then said he thought it was very interesting which was Bohr language for saying that it wasn't. If he had been really interested, he would have engaged in a Socratic dialogue. Living a life as I said is not like crossing a field.

## A Technical Appendix

In this appendix, I want to fill in some of the technical details of the suggested BLYP experiment. We are only interested in negative muons with the decay mode $\mu^- \to e^- + \nu_\mu + \nu_e$. The lifetime for this decay if the muon is free is about 2.2 microseconds. We are interested in muons bound in mu-mesic atoms. The lifetime will be different but we will neglect this relatively small effect.[5] Let me call the free decay rate d. In the absence of capture, at any given time the number of muons is given by

$$N(t)_\mu = N(0)_\mu e^{-dt}. \tag{1}$$

But the total number of muons plus electrons is conserved and equal to $N(0)$.
Thus

$$N(0)_\mu = N(0)_\mu e^{-dt} + N(t)_e. \tag{2}$$

Or

$$N(t)_e = N(0)_\mu (1 - e^{-dt}). \tag{3}$$

Next I want to consider the case in which the muon is captured by a proton at the rate R; the reaction $\mu^- + p \to n + \nu_\mu$. In this reaction the muon disappears and no electron results. As mentioned above, we considered the simplified model of the nucleus, in which there was a single proton that carried the angular momentum I around a core of Z–1 protons, which have no spin. There are two hyperfine states with total angular momentum $I + \frac{1}{2}$ and $I - \frac{1}{2}$. Because of the spin dependence of the capture mechanism, the capture rates from these two states $\lambda_+$ and $\lambda_-$ are different. There can also be a rapid transition, from the hyperfine state of higher energy to the lower one. One must also know how these states are initially populated.[6] The precise values of the capture rate also depend on one's assumptions about the weak interactions. I will not work these out here but refer to our paper, noting once again that in light of our present knowledge of the V,A nature of the weak interactions, the formulae we gave are more complicated than they need to be because we took all the possibilities, such as the tensor coupling, into account.[7] If we allow capture, then Eq. (1) must be replaced by $Ce^{-dt}(1-Ae^{-Rt})$. Here R is the capture rate and A is a parameter determined by $(\lambda_+ - \lambda_-)/\lambda_{avg}$. The pluses and minuses in the two lambdas refer to the two hyperfine capture rates and $\lambda_{avg}$ the average capture rate including the Z–1 protons in the core. For Z greater than 1, this says that the hyperfine effect goes a 1/Z, so to measure it one must use relatively light nuclei. It was first successfully measured in an isotope of fluorine. The rate R depends on the hyperfine state. As the time evolves the muons will transition to the lower hyperfine state so R is effectively $\lambda_-$. For Lee and Yang to have seen all this and to have written it up over a long weekend was for me nothing short of a miracle.

## Endnotes

1. The actual paper was published in *Phys. Rev.* **111**, 313 (1958).
2. The quantum interference depends in $e^{i\Delta vt}$ and if the argument of the exponent is large the term oscillates to zero which is the usual way decoherence is established.
3. See *Phys. Rev.* **129**, 2766, (1963) By the time of the publication of this paper the weak interaction was known to be vector and axial vector which maximizes the effect. The analysis in this paper is much more detailed than in ours and the experimental results for LiF are given. The results are consistent with V–xA where x is 1.21. As discussed in the appendix the relevant parameter for characterizing the effect is the difference of the capture rates in the two hyperfine states divided by the total capture rate including the core of Z–1 protons. For large Z this ratio goes as 1/Z hence one does the experiment for relatively small Z.

4. H. Primakoff, *Rev. Mod. Phys*, **31**, 802 (1959)
5. There are a number of effects that must be considered. These include relativistic effects such as time dilation and the use of a bound state wave function. A relatively straight forward one is due to the effect on the muon mass due to its Coulomb binding. If G is the Fermi constant and m the muon mass then dimensional analysis shows that the decay rate is proportional to $G^2m^5$. The mass is reduced by a factor proportional to $(Z\alpha)^2$ and hence the effect is small but contributes to the reduction of the decay rate.
6. An early treatment of this can be found in R.Winston and V. L. Telegdi, *Phys. Rev. Lett*, **7**, 104 (1961). This has been much discussed in the subsequent literature.
7. One must distinguish between what I would refer to as "intrinsic" and "induced" couplings. The former are in the nature of the force itself and the latter are "induced" in the course of perturbative calculations. We considered only the former. An example of the latter is a negative muon emitting a virtual negative pion and a neutrino. The proton absorbs the pion becoming a neutron. The appears as a pseudo-scalar in the capture rate. There is now a vast literature on these effects and their experimental consequences. It is beyond the scope of this brief note to comment on it.

# Murray 4

Murray

Columbia Grammar School Yearbook 1944

I first heard, or more exactly, read, the name Murray Gell-Mann in the spring of 1944. It appeared in the yearbook of the Columbia Grammar School. The school was, and is, located on the west side of Manhattan near Central Park. At the time, most of the students were

Jewish. A few like Murray were there on scholarships and the rest paid tuition. In the spring of 1944 I was finishing my freshman year and Murray was about to graduate. We were nearly the same age — fifteen — Murray being a few months older. He was born on September 15, 1929 and I was born on December 31. He had already been admitted to Yale.

Looking at the yearbook I find a solemn picture of Murray with a long list of his extra-curricular activities, ranging from the soccer team to the debating club. There was a page called Senior Statistics. Murray was voted as the most studious. "In a shower of votes, Murray Gell-Mann the "wonder boy" easily won the title." Under Prophecy a wag wrote, that when a bomb had been hurled by the Soviet Embassy to America "Gell-Mann had forgotten to light the fuse." Murray appears in no other category including "Most Likely to Succeed." While, I do not remember laying eyes on Murray during our overlap, I vividly remember one of the traces he left. We shared in different years, a somewhat vitriolic math teacher — one James Reynolds. Red in the face, mister Reynolds would compare us with some of the giants he had taught in the past, including Gell-Mann and ending up by saying that in Tarrytown, where he lived, "we bury our dead." Murray had a favorite teacher, as did I, Dow Bunyon Beene. Beene I think had been a minister in the south and now he taught subjects like English and Latin. He was vivid and scatological — a wonderful teacher for boys. Over the years I have noted that Murray made contributions to a school fund in his name. He took the same high school physics course I later did and found it abominable.

I don't know what it was like to be fifteen and a Jewish scholarship student at Yale. I would imagine it was not entirely easy. Murray majored in physics and was required to write a senior thesis on a subject he was assigned. Murray had no idea how to tackle the subject and never wrote the thesis. Not surprisingly when he applied to Yale for graduate work in physics he was turned down although he was admitted to the math department. He applied elsewhere such as Princeton and was either turned down or not offered the scholarship he needed. Then a sort of miracle happened. Victor Weisskopf — Vicki to one and all — who was the leading theoretical physicist at M.I.T. offered Murray the job of being his assistant. I have never been able to learn how this happened but Vicki was the perfect person for Murray to have done a thesis with. He was interested in all of physics and had wide cultural interests as well. He also had a very good sense of humor and had seen many geniuses come and go. He gave Murray a problem in theoretical nuclear physics and Murray came up with some original work that he could have published, but did not bother. It was re-discovered by others. He must have made a very favorable impression on Vicky, who in 1951, helped to get him admitted to the Institute for Advanced Study in Princeton. He was twenty-one. Murray often tried out new ideas on Weisskopf. When he tried out the quark with its bizarre electric charges, Weisskopf asked him if he had been smoking anything.

At the time, and even a few years later when I was admitted, the supply of offices was limited. To deal with this we were assigned office mates. In my first year, I was assigned a young Italian physicist and in my second year, I was assigned a Chinese American who later became a professor at Harvard. Murray was assigned Francis Low. It was the most fortunate choice imaginable. Francis was then thirty and he later remarked that he had found this child in his office. During the war, Francis had volunteered for the ski troops and had served in Italy. He had then gone to Columbia for his graduate work. Francis was one of the most brilliant theoretical physicists of his generation. He was very fast and extremely clear. Murray had met his match. The two of them wrote a paper which is still regarded as a classic. The strength of the interaction between two quantum fields is indicated by a number called the "coupling constant." As they pointed out it is not really a constant, but depends on the energy of the interaction. They produced an equation describing this dependence. It turns out to have applications in several areas of physics. Francis told me that Murray tended to take over the seminars. It must have reminded Robert Oppenheimer who ran them of his own early days in Gottingen when he did the same thing.

After the Institute Murray had a couple of temporary jobs at the University of Illinois and Columbia. In 1954, he went to the University of Chicago as an associate professor. Enrico Fermi had been the dominant figure at Chicago and indeed in the entire mid-west. But in 1954, he died and the theoretical physics of the area kind of imploded. Francis who had been at the University of Illinois went to MIT and Murray went to Cal Tech. The dominant figure at Cal Tech was Richard Feynman. I do not think that their relationship was ever entirely comfortable. A colleague of mine witnessed Murray's first seminar at Cal Tech. He said that Feynman insisted on knowing why Murray had hyphenated his name. It is hyphenated in my year book. I did once meet his older brother Ben who told me in no uncertain terms that he did not hyphenate his name and spelled it Gelman. I later witnessed a scene between Murray and Feynman. It was at a small conference in Gatlinburg Tennessee. We had our meals at community tables. As luck would have it, Feynman sat at the head of our table. Murray came over and began talking to Feynman, as if he was some sort of graduate student. When he left, Feynman commented that Murray was not used to talking to people who were smarter than he was. I am sure that Feynman found some of Murray's non-physics interests somewhat silly. Murray was an avid bird watcher and went all over the world to see new birds. He was also a serious linguist with an extraordinary capacity to learn languages. A friend of mine went with Murray to Mexico on a trip. At some slightly remote location they encountered a Mexican, for whom Spanish was a second language after some local Indian dialect. Murray managed to carry on a conversation with him in his language.

Until the spring of 1959, I do not recall having a conversation with Murray. Then three things happened. Murray wrote a short paper on a subject that I knew something about. It was an idea that cried out for experimental testing. A colleague at the Institute and I wrote a paper in which we made a systematic study of the possibilities. About the same time, I heard that I had won a two-year National Science Foundation fellowship which would enable me to travel anywhere to use it. I had chosen Paris and indeed had made some arrangements with a French physicist named Louis Michel, whom I had met while he was visiting Princeton. The third thing that happened was that Murray appeared. He showed up in my office and asked what I was going to do the next year. I said I was going to Paris. He said so was he and added "Stick with me kid, I will put you on Broadway." I don't think we discussed Columbia Grammar. Michel was then a professor at the Ecole Polytechique, a military school that he had attended which was then located in the Latin Quarter in Paris. Michel said he would find me someplace to stay, which turned out to be a fairly primitive studio in the Cuban House at the Cite Universitaire. Michel had his office in what seemed to be the basement of a building at the school. I just took another desk.

I had come to Paris with a physics problem that I had not been able to solve. I explained it to Michel and suggested that we work on it together. He was enthusiastic so that first afternoon, when I was in our office, we began to discuss it. Much to our surprise Murray appeared. He was then a visiting professor at the College de France. How he found us I will never know. Murray asked what we were working on and I told him. He left and Michele and I continued our discussion. The next afternoon he re-appeared and announced that le problème est résolu— the problem is solved. He left the draft of something and disappeared. Michele and I discussed what we should do. I suggested that we keep working on it in case we came up with something different. Somehow Murray decided that what he had done was somehow not quite right and suggested that we write a joint paper, which is what happened. The problem was mine but the solution was Murray's and I never would have come up with it. Murray's French deserves a comment. It was perfect — maybe too perfect. He would translate into French, English terms that the French never translated. There was something that appeared frequently called gamma 5. Murray always called it gamma cinque, which no French physicist would have done.

Murray was in Paris with his wife Margaret Dow, a British-born archeologist whom he had met at the Institute and their very young daughter Lisa. I went on some walks with Lisa, who showed an uncanny ability to name the brands and model years of French automobiles. I will come back shortly to what Murray was primarily working on while he was in Paris but I want to mention his trip to Israel which accidentally coincided with my own. Murray always referred to Israel as "occupied Palestine" so I was surprised the previous spring, when he asked to accompany me on a visit I was making to Abba Eban in Washington, who was then Israel's ambassador to the United States and a friend of my father's. I was a little embarrassed when Eban asked Murray whether he too was a physicist. The Israelis were courting Murray, so they laid out the red carpet which included an ornithologist, an alleged expert on the birds of Israel. I was amused when after being grilled by Murray, the poor fellow tried to change the subject, by asking Murray some questions about elementary particles. I also went with Murray on a notable trip to the Negev desert to visit David Ben-Gurion. Ben-Gurion was a short, very lively man. At one point he pointed to a barren hill and asked Murray if he could see the trees. Murray looked puzzled and Ben Gurion told him that if he came back in ten years there would be trees. On the way back to Tel Aviv, we took a wrong turn and ended up in front of the reactor, which was making the plutonium which the Israelis were using for their nuclear weapons. We beat a hasty retreat.

When it came to physics, I found it very hard to keep up with Murray. I would hide out. I remember at one point getting a phone call from him, asking where I had been. "I have discovered millions of things," he explained. When I saw him, he said that he had discovered that the currents he was studying commuted like angular momenta. I could see that this was true, but I had no idea why it mattered. When he got back to Cal Tech in the fall he discussed this with a mathematician, who explained what he was trying to do. This led to the great paper by Murray, where he created the mathematical symmetries of the elementary particles, some of which had not yet been observed. When they were, Murray won the Nobel Prize in 1969. There have been some strange Nobel Prizes but this one wasn't.

The Aspen Center for Physics was founded in 1962. It was to be a place where physicists could come and do research, without any of the distractions of the usual academic life, such as students. Murray was an enthusiastic participant. He even bought a lovely old house not far from the Center. I do not think Murray was a mountain climber, but he was an enthusiastic hiker. I remember meeting him and his family on a trail. He asked me if I had seen any interesting birds. When I told him that I had seen a couple of small grey ones, he looked at me in disgust. I once visited him in his office at Cal Tech in Pasadena. On the blackboard, there was a notation which read something like Crawford Greenewalt — hummingbirds. Crawford Greenewalt was then president of Dupont and was asking Murray to explain the remarkable colors of hummingbirds. After a terrible bout with cancer, his wife Margaret died in 1981. She is buried in Aspen.

It was clear when I had a chance to talk to Murray in Aspen that his interests were no longer focused only or even primarily on physics. He was studying the complexity of systems and was in demand for various things such as the McArthur foundation. In 1984, he helped to found the Santa Fe Institute and left Cal Tech for New Mexico. He had written a successful popular book, The Quark and the Jaguar. When I look at our 1944 Columbia Grammar yearbook, especially at the members of the senior class who were voted most likely to succeed, I would imagine that Murray was the last person his classmates would have voted for, but he was the only one who really did. The rest are forgotten.

# II Chronicles

# 5 Houtermans

First Published December 8 2016, The New York Review of Books

In the 1930s German and Russian scientists of Jewish origin were treated quite differently. The German scientists were automatically "guilty" if they had more than one eighth Jewish "blood." In Russia they had to be guilty of doing something. Until October 1941 Jews were encouraged to leave Germany and most of the prominent scientists did so, though not without great difficulty, since they were allowed to take very little money out of the country. The Russian scientists, on the other hand, were not allowed to leave. Many were arrested and forced to "confess" to crimes. After having done so they were either imprisoned or summarily executed. In both cases the amount of scientific talent destroyed was beyond calculation. Here are two examples from German science.

Herman Mark was born in Vienna in 1895, making him an Austrian citizen. His father had been born Jewish but converted to Lutheranism upon marriage. A highly decorated World War I veteran, Mark became a very distinguished polymer chemist and was at the IG Farben company in Germany until the Nazis came to power in 1933, when he returned with his family to Vienna. But after the German annexation of

Austria in 1938 he was arrested by the Gestapo. He was released without his passport, which he retrieved with a bribe of almost a year's salary. He then left on a "ski trip" to Switzerland with his family, in a car with skis on the roof and a Nazi flag on the hood. He had bought some $50,000 worth of platinum wire that he fashioned into coat hangers. His wife had knitted covers for them. He ended up at the Polytechnic Institute in Brooklyn, where he assembled a noted group of chemists. His son Hans, who was on this "ski trip," became a physicist and served as the secretary of the US Air Force.

The nuclear physicist Hans Bethe is another example. He had two Jewish grandparents, so according to the first racial laws that were promulgated in 1933 he could not hold a university position. He was at the time at the University of Tübingen and was immediately dismissed. His former professor Arnold Sommerfeld — a very great teacher of physics — made strong efforts to find jobs abroad for his Jewish students. Bethe found a temporary job in Manchester and then got an offer from Cornell, where he spent the rest of his career. He won the Nobel Prize in 1967. During the war he was head of the Theoretical Division at Los Alamos. Late in his life he said he regretted that the Manhattan Project had not made the bomb earlier so that it could have been used against Germany.

Much has been written about the fate of Jewish scientists in Soviet Russia, but I have never seen anything quite like *Physics in a Mad World*. It consists of two long essays on the physicists Friedrich Houtermans and Yuri Golfand. Both of these are written by Russians who had access to police files. The book's editor, Mikhail Shifman, now a professor at the University of Minnesota and a distinguished theoretical physicist, was also born in Soviet Latvia. He supplies useful comments. The essay on Houtermans by Viktor Frenkel, which takes up much the largest portion of the book, is particularly remarkable, but Golfand also deserves close attention.

Golfand was born in 1922 in Kharkov in Ukraine. His father was a nonobservant Jew who was an engineer in the meat business. Yuri showed mathematical ability at a young age. In 1939 he became a student at Kharkov University and was then sent to the Air Force Engineering Academy, where he reluctantly became a military engineer, which saved him from going to the front. After the war there was an outburst of anti-Semitism, so he could not get into any university to study for a graduate degree. He worked in the Moscow Computing Center and started graduate school by correspondence.

Much later, after receiving a master's degree by correspondence, he managed to get a job at the prestigious Physical Institute of the Academy of Sciences (FIAN) in Moscow. One of the groups there worked on the design of the Russian hydrogen bomb. Golfand did computations for the construction of the bomb without knowing what they were for. Later, his experience at FIAN made it difficult for him to leave Russia because it was claimed that he knew military secrets. He became what was known as a "refusenik" since he was denied an exit visa.

In 1971 Golfand and an assistant published a paper on "supersymmetry" that became one of the cornerstones of modern elementary particle theory. In nature there are two classes of particles — bosons and fermions. An example of the former is the Higgs boson and of the latter the neutrino. Supersymmetry relates the two classes and predicts that for every boson there is an as yet undiscovered fermion and vice versa. If this turns out to be true, then it would help to explain why the various forces are conjoined in a unified scheme. Golfand was one of the first to suggest this idea. The Large Hadron Collider, built by the European Organization for Nuclear Research near Geneva, is trying to produce these particles.

In 1972 Golfand lost his job at FIAN because he was Jewish. He began trying to emigrate to Israel in 1973 but it took him seventeen years to get permission to leave. He was "invited" by the KGB to visit its headquarters on several occasions and always refused to show up. The miracle is that he was never thrown in jail and somehow kept working at home. His wife took menial jobs to support them. He was eventually offered an academic job in Russia in 1980 but he wanted to get out.

He finished his career at the Technion in Israel and died there at the age of seventy-two. In view of what he did manage to accomplish under extremely difficult circumstances, one wonders what he could have achieved if he had been allowed to flourish in the Soviet Union.

Over the years I have read bits and pieces about Friedrich Houtermans but nothing like what is in *Physics in a Mad World*. The long profile was written by Victor Frenkel, a noted biographer and historian of Soviet science who died in 1997. Shifman has filled in the gaps. If there were a movie of Houtermans's life, no one would believe it.

Houtermans was born in 1903, in Zoppot near Danzig. His mother was a Viennese, half Jewish, and the first woman in Vienna to receive a doctorate in biology. His father was Dutch and quite well to do. When Houtermans was three his parents divorced, and he moved to Vienna with his mother, who became a prominent cultural figure there. She decided that young Houtermans could use some psychiatric counseling so she engaged Freud, who told him to leave after it became clear that Houtermans was making up his dreams.

Houtermans attended a prestigious gymnasium and excelled in mathematics. He also became interested in *The Communist Manifesto*, which he read aloud to his classmates — something his teachers did not appreciate. In 1922 he enrolled in the University of Göttingen, one of the greatest centers for physics and mathematics in the world. A few years later he wrote his Ph.D. thesis there.

Another student at Göttingen was Robert Oppenheimer. On a train trip — a visit to Hamburg — all the students except Oppenheimer had extremely shabby luggage. The one woman in the group, Charlotte Riefenstal (no relation to Leni), admired Oppenheimer's pigskin suitcase. Sometime later Oppenheimer appeared at her doorstep with it and in a gesture of courtship gave it to her. She kept the suitcase but married Houtermans, becoming his first and then third wife. He moved on to Berlin where he made a very important contribution to astrophysics with the British astronomer Robert Atkinson. This was based on work he had done before he moved to Berlin with the Russian physicist George Gamow, whom he had met in Göttingen.

Göttingen was one of the centers for the development of the then new quantum mechanics. Max Born was an immensely respected senior professor. His assistant Werner Heisenberg invented a version called "matrix mechanics." Heisenberg did not know what a matrix was until Born explained to him that he had in fact created one. Heisenberg found that in his expressions the quantities did not obey the normal laws of arithmetic. For example, AB was not equal to BA. He just accepted this without knowing what it meant. Born realized that it meant that A and B were matrices — arrays of numbers — and that these matrices do not necessarily obey the AB=BA rule. Thus a form of quantum mechanics was created that involved matrices. In the meantime, Schrödinger found another form that involved ordinary equations. It was soon realized that these were simply different expressions of the underlying theory.

Gamow made one of the first applications of quantum mechanics to nuclear physics. Some nuclei decay into so-called alpha particles, which are actually helium nuclei. These nuclei have a positive charge so it was not clear how they escape from the mother nucleus, which has a positively charged barrier. Classical physics said this was impossible because the nucleus, it was thought, would repel the alpha particle and it could never get out, but in quantum mechanics the alpha particles can tunnel through the barrier. This is what Gamow worked out.

He and Houtermans wrote a paper on some of these processes. When Houtermans moved to Berlin, he and Atkinson considered the possibility that the barrier could be penetrated from the outside so that two light nuclei could fuse. Because of the difference in the masses of the two nuclei, energy is produced. They realized that the conditions in stars — including the sun — are such that fusion can take place and that this can account for stellar energy. This is probably the most important contribution Houtermans made to physics, but he made many others.

In the 1920s Houtermans joined the German Communist Party. This, along with his Jewishness, meant that he had to get out of Germany in the 1930s. He and Charlotte first went to England, where he did research for the EMI Television Laboratory. In 1935, he chose to go east rather than west, which nearly cost him his life. He joined the Ukrainian Physico-Technical Institute in Kharkov, where he had excellent colleagues. By 1937 an official Soviet declaration stated that "enemies [had] penetrated among the physicists, carrying out espionage and sabotage assignments in our research institutions," and that it was the duty of physicists to unite "around our great leader Comrade Stalin." From that point on, it was only a matter of time before Houtermans would be arrested. This happened in the first few days of December 1937 in Moscow.

*Physics in a Mad World* provides the details of his interrogation by the KGB. For days on end he was kept awake and standing. If he fell asleep he was doused with ice water. His feet swelled to the point where he could no longer wear his shoes. He was finally broken when he was told that his wife was going to be arrested, that his children were going to be put in an orphanage under new names, and that he would never see them again. Actually Charlotte and the children had already escaped from Russia. There is a detailed description in the book of how they did this, partly with the help of Niels Bohr. Charlotte eventually ended up in the United States and taught at Vassar.

Houtermans knew nothing of their escape. He was put back in prison. In Shifman's book there is a stunning description of him by a fellow prisoner, Konstantin Shteppa, professor of history at the University of Kiev. Shteppa had just entered Houtermans's cell:

I entered the cell which contained a single piece of furniture a wooden bunk-bed. Immediately I was shocked. On the top bed laid a corpse. The man's face was grey and the skin was so thin that one could see every bone under it. I was terrified. "Is it possible that they've become so cruel? That the degree of their mockery has reached the point of putting the dead and the living together?" That was my first thought after I saw his face.

After a while he opened his eyes. He stared at me with a look of expectation, hopeful I would bring him all kinds of news, which he needed so desperately.

"Are you new? I can tell by the way you walk, you move and you look," he said in his broken Russian. He lifted himself up and offered his thin hand for a handshake. "My name is Fritz Houtermans, a German...a physicist...a former member of the communist party...former emigrant from Nazi Germany...former professor at the Institute of Physics in Kharkov...former human being — and who are you?"

Houtermans asked for cigarettes, which Shteppa did not have. Houtermans was a lifelong chain smoker and eventually died of lung cancer. He kept himself sane by doing number theory in his head. He found a proof that the equation $a^3+b^3=c^3$ has no solution for nonvanishing integers. This is not trivial. The great eighteenth-century mathematician Leonhard Euler found a proof that was somewhat faulty. Houtermans describes his attempt. First he tried to use matches to write on soap, but that did not work. He writes:

When I found on August 6th an elementary proof for Fermat's famous problem for $n=3$, which I have learned since is essentially the same as Euler's, by "descent infinite,"[1] I got very excited about it, because I did not know Euler's elementary proof to exist, and I applied to the People's Commissar of the Ukraine to get paper and pencil. (I said I wanted to work out an idea of mine on a method of radioactivity which might be of economic importance.)

When my petition was not granted, I went on a hunger strike (only declining food, not water). I was alone in the cell then and succeeded in getting pencil and paper after 8 days of hunger strike, by which time I was very much weakened since I had been in a bad state when I started. I wrote a number of theorems.

Later he published some of the theorems he had written out in prison.

The Soviet–Nazi 1939 nonaggression pact changed his circumstances. There was an exchange of prisoners and Houtermans was sent to Germany, where he was immediately arrested by the Gestapo and ended up in a prison in Berlin. His cellmate was about to be released, so Houtermans wrote a brief note stating his whereabouts, which he asked to be delivered to some of his former colleagues. One of these people was Max von Laue, a Nobel Prize–winning physicist and an outspoken critic of the Nazis. It seems a miracle that he was not put in jail. Because of his prestige he not only got Houtermans out, he also found him a job.

Manfred von Ardenne was a German aristocrat who had become very wealthy because of his electronic inventions. He had an estate outside Berlin where he maintained a substantial laboratory funded by the post office. He could hire anyone he wanted, and he hired Houtermans. I have read Ardenne's autobiography and I cannot separate fact from fiction in it. There is no question that at the time Houtermans joined the laboratory, the scientists there were doing nuclear physics. To what end is still not clear, but looked at from the outside they were certainly studying nuclear energy and its applications. There was also a rival official program that Heisenberg eventually headed and that tried to ignore von Ardenne's laboratory.

Both the official program and von Ardenne's laboratory hit on the idea of using element 94, which its American discoverers named plutonium, for nuclear energy — a fact that was concealed until after the war. This was Houtermans's discovery. He was so alarmed by it that he tried to inform some of his colleagues in America. A communication from him did arrive but no one knew what to make of it. When the Russians occupied their part of Berlin they seemed to know about Ardenne's laboratory. Ardenne went willingly to Russia and headed a laboratory there that worked on the Russian atomic bomb program. Houtermans was long gone.

His first destination was the Imperial Physical-Technical Institute in Berlin. In 1943, he took advantage of a German law to divorce Charlotte, who was still in America, on the grounds that they had not lived together for many years. He remarried and had three children with his new wife. Then he went to the United States and remarried Charlotte, whom he divorced again after two years. He soon became a professor in Bern and married again. By this time he had changed his field to earth sciences. He invented new and better ways of measuring the age of the earth. He died in Bern in 1966. His creativity under terrible circumstances was almost incredible. One can only imagine what he could have done if he had been left alone.

**Endnotes**

1. "Descent infinite" refers to a method of proof that proceeds as follows. First suppose that you have found three nonvanishing integers that satisfy $a^3 + b^3 = c^3$. (If one of the integers vanishes then the solution is simply that the cube of one equals the cube of the other, which is obvious.) Suppose you can show that if this is true then there must be three smaller ones that also satisfy the equation. This part is not trivial. But then there must be three still smaller ones so there can be no smallest one, which is a nonsensical outcome. This provides a proof by contradiction. One has assumed an integer solution to the cubic equation and has shown that this leads to a nonsense.

## Charlotte 6

Göttingen-Riefenstahl in the front row. The 65th birthday of Gustav Tammann director of the Göttingen physico chemical laboratory.

Charlotte Riefenstahl (no relation to Leni) was a very attractive woman. She was also nearly unique for her era, the 1920's in Europe. She was a physicist. There were a few others at the time such as Marie Curie and Lise Meitner. To give an idea, when Meitner, who later became the first person to correctly analyze nuclear fission, came from Vienna to Berlin to attend the lectures of the great German physicist Max Planck at the Friedrich Wilhelms Universität, she was the only woman whom Planck had ever allowed to attend his lectures. When she became his assistant there was no women's restroom in the building. She had to go through the laboratory and exit the building to use the facilities of a restaurant across the street. This was the early 1900s. Riefenstahl did not face quite this kind of prejudice two decades later when she did her studies in Germany. We know how she felt because she wrote about it. The various fragments have been put together in an autobiographical essay, which introduces the collection, Standing Together in Troubled Times, edited by the physicist Mikhail Shifman, who also edited the book Physics in a Mad World. Shifman has managed to find many previously unpublished letters to and from Riefenstahl, involving some of the greatest physicists of the twentieth century.

Riefenstahl was born on the 24th of May 1899 in Bielefeld, Germany. Her father was a journalist who managed to earn enough money to send her to a good school. She was in training to become a high school teacher, which she did after her father died, in order to help support her mother. When she learned that her mother was actually saving this money for her bother Karl's education, she decided instead to save this money for her own education. When she had enough, she decided to go to Göttingen, which was arguably at the time the finest place in the world to study physics and mathematics. On the mathematics faculty was David Hilbert, from whom she took courses, who is often considered to have been the greatest mathematician of the twentieth century. On the physics faculty was an array of future Nobel Prize winners, including Max Born. She stayed there from 1922 to 1929, when she got her PhD. During that period the quantum theory was invented in the first instance by Werner Heisenberg, who was Born's assistant. Born himself invented the probabilistic interpretation of the theory. Among the students was P.A.M Dirac who invented his own version of the theory and Robert Oppenheimer. They were good friends and Oppenheimer, who wrote poetry, often told of an exchange he had with Dirac about this. Dirac said, "In physics we take a complicated situation and try to describe it in the simplest possible terms, whereas in poetry..." Oppenheimer had a "crush" on Charlotte. I think there is no other term better suited to describe his somewhat adolescent behavior, which consisted among other things of his giving her an expensive suitcase of his which she had admired. She writes

"I knew that Robert was attracted to me. Women feel such things. He courted me as best he could in his stiff and excessively polite way. He seemed strange to me, and far away, as far as a distant nebula. Or perhaps, I unappreciated him at that time. I saw his gentleness, courtesy and humor, but he was very un-European. Much later he himself said, 'Although [my Göttingen friends] were warm and helpful to me, they were packed there in a very mysterious German mood...bitter, sullen and I would say discontent...'"

Her romantic life changed radically, with the appearance in Göttingen of Fritz Houtermans, known to all his friends as Fisl. She writes,

"Fisl was a completely different story. He was very Viennese and very confident of himself. When Fisl appeared in Göttingen in 1921 after having earned a baccalaureate in Wickersdorf he behaved and also looked quite different from everybody else who had gone through a regular school. With his blue eyes and dark hair, which was cut in Viennese fashion, [As far as I can tell this meant leaving a good deal of hair on top.], his lanky behavior and slight stoop, he was different from the ordinary run of young German academics, as if he had come from the moon. I always thought he looked Italian..." They met in the winter of 1926–27.

The remarkable life of Houtermans was one of the subjects of Shifman's previous book. Charlotte plays only a relatively minor supporting role. Here she is front and center. What struck me is that although they were not married, they seem to have lived together, something that I don't think was common at the time. Prior to their meeting, she had decided that to earn a living, she had better get a teaching job and she thought getting one in the United States would be a good idea. There was no question of getting such a job anywhere but a woman's college. Women in the sciences were, in the main, second class citizens although no one would have dared to describe Marie Curie this way. Riefenstahl describes the case of Maria Goppert, who was a younger contemporary at Göttingen. She married one of the assistants, the American physicist Joseph Mayer. Her husband got a job as an assistant professor at Johns Hopkins whereas the only job she could get was as a department secretary. This inequality persisted when he moved to the University of Chicago, where she became an unpaid associate professor. This lasted until close to 1963 when she won the Nobel Prize. Charlotte got a temporary job teaching physics at Vassar for a year and then for a year at Winthrop College in South Carolina. These experiences turned out to be crucial for what happened to her later. After the two years, she returned to Germany and to Fisl. Incidentally when she arrived at New York, she was met by Oppenheimer who had come to the pier in his father's chauffeur-driven limousine. She was taken to the grand Oppenheimer apartment on Riverside Drive in Manhattan. On the walls were Picassos and Van Goghs. One of the latter was on the wall of Oppenheimer's house in Princeton where I saw it in the fall of 1957.

In 1930, she and Fisl got an invitation to attend a conference in Russia. Fisl was a member of the German communist party and welcomed the visit. They were taken on a boat trip to the Crimea and as she tells it on the deck, Fisl proposed to her and she accepted. They eventually had two children together. It soon became clear that life in Germany was becoming intolerable. Fisl had enough Jewish ancestry so that the racial laws applied to him. They went to England where he worked in industry for a while. They did not like it much and Fisl had an offer from the Ukrainian Physics and Technology Institute. Despite warnings from colleagues, he took it and they took up residence in the Soviet Union. Fisl got swept up in the arrests of intellectuals and nearly died in Soviet prisons, until he was finally sent back to Germany where he was arrested by the Gestapo. A few of his former German colleagues managed to get him released and he had a fairly comfortable war, working for a private laboratory. Meanwhile Charlotte had been allowed to leave and went temporarily to Copenhagen where she and her two children were looked after by Niels Bohr. He gave her enough money to get by for a while. Indeed, he even managed to get her a special League of Nations passport as her German passport had expired. There is a section of the book where she describes her harrowing train trip from Russia to Copenhagen. It was a very close call and it was lucky she made it. After Copenhagen she went to England with the children where she was nearly destitute. In her desperation she wanted to ask the Oppenheimers for some money. One of the most remarkable letters in the book is one from the physicist Wolfgang Pauli to her about Oppenheimer. Pauli was one of the great physicists of the twentieth century and was also noted for his acerbic criticism of scientific work. But he was also a very shrewd judge of people. Here is some of what he wrote to her about Oppenheimer whom he knew very well.

*"I pondered about Robert Oppenheimer: he is certainly very irrational towards women. After some time, he always shows tendencies to escape whenever a human relationship of any kind (not only love) develops with a woman. (If his corresponding sexual inhibitions towards women are the cause of these escape tendencies or only their consequence is difficult to say.) So it seems to me that he wants to elude you, but I do not believe at all that money plays any role for him in this connection. So it is maybe the best that you just send him a note asking whether he received your letter. But maybe something has already arrived from him in the meantime."*

She did receive a letter from him in which he noted that his father had recently died. Enclosed was a check for fifty dollars. Not long after, she and Pauli learned that he had actually gotten married. He had inherited one of Van Goghs which was mounted on the wall of his living room in Princeton. I don't have the impression that Charlotte had much, if any, further relationship with the Oppenheimers even after she moved to the United States.

After a brief stay in England Charlotte came to the United States, where she took up teaching jobs in various women's colleges such as Radcliffe and Wellesley until she finally ended up in Sarah Lawrence where she taught for over two decades. As soon as she was free of Russia, her priority was to liberate Fisl. She corresponded tirelessly with celebrated physicists such as Einstein and even exchanged letters with Eleanor Roosevelt, some of which are in the book. In 1941, she received word that Fisl was in Berlin and alive. She writes, "My first thought was 'Thank God, Fisl is alive and well! We will reunite.' It took an hour or so to realize that most probably he was in the hands of the Gestapo. For the first time in two years I knew he was alive and well." In 1942–43, Houtermans published three papers with a collaborator Ilse Bartz. Charlotte learned about them from an old friend who had been to Göttingen with them. What she did not know at first was that Houtermans had fallen in love with her. Indeed, he was able to file for divorce on the grounds in German law that a husband could divorce if he had not had conjugal relations for a certain time. One can imagine her feelings, which she vividly expresses in the book. That marriage did not last and remarkably Houtermans, who had moved to Switzerland, proposed that the two of them remarry. Even more remarkably she consented. But she found that Houtermans was drinking heavily and chasing other women so that marriage only lasted six months, after which she returned to the United States where she spent the rest of her life. She died in 1993 in Northfield, Minnesota. She had moved there after her retirement to be near her daughter.

In reading this book, I realized that I owed her a debt. When I was a student I wanted to teach myself about the quantum theory of fields. There was one recommended text, Quantum Theory of Fields, written in German by the German-born Swiss physicist Gregor Wentzel. I was able to read the English translation, the one that she had made.

# Einstein and the Fraud 7

Albert Einstein and the Fraud

Philipp Heinrich *Emil Rupp*

In 1932, the *Zeitschrift für Physik*, one of the most prestigious physics journals in the world, published a brief statement by the physician E. Freiherr von Gebsattel. It is certainly the strangest bit of writing ever to appear in a physics journal. It reads in part,

"Dr. [Emil] Rupp has been ill since 1932 with an emotional weakness (psychasthenia) [This is an ill-defined term that is not used anymore but encompasses a great variety of psychic disorders.] linked to psychogenic semi consciousness. During this illness, and under its influence, he has, without being himself conscious of it published papers on physical phenomena... that have the character of "fictions". It is a matter of dreamlike states into the area of scientific activity."

There are few things that are left out of this extraordinary statement. Rupp had been publishing physics papers since the 1920's and nearly every one of them was fraudulent. More remarkably many were done with the active collaboration of Albert Einstein.

Mistakes in physics have appeared in print. For example, Enrico Fermi was awarded the Nobel Prize for Physics in 1938 in part for discovery of elements heavier than uranium. In fact without knowing it, he had instead discovered nuclear fission. There was no fraud here. He just made a mistake. I also think that there was no fraud in Joseph Weber's claim that he had detected gravitational waves. This was also a mistake. In both cases these were fairly rapidly corrected by the community. Now nearly all experiments are done by groups. The Higgs boson for example, was discovered by two teams totalling some 3000 people. Perhaps a professor works with a few graduate students and some technicians. All of them would need to be complicit in a fraud. If the discovery was interesting, the experiment would be repeated elsewhere, which is what happened to Fermi and Weber. Two very recent examples are the "discovery" of what is known as B Mode Polarization in the Cosmic Background radiation and the "discovery" of a supermassive Higgs. Both of these made headlines. The former would have shown the effects of gravitational fluctuations in the very early universe and the latter a new domain of particles. Both of these turned out to have been mistakes. Not fraud but mistakes. As I will now explain, Rupp exposed himself.

Rupp was born in 1898. He got his PhD in Heidelberg and remained there for a while to do experiments on the radiation emitted by so called "canal rays" — charged particles that are accelerated in a tube. In the mid-1920's, when Rupp began his work, the quantum character of radiation was not well-understood. In 1926 Einstein decided, having read some of Rupp's work, his techniques could be adapted to learning about this and he proposed a specific experiment which Rupp began to perform. When he sent some of his results to Einstein, Einstein would point out flaws and in the next communication Rupp would claim to have corrected them, leading to more flaws. It is unclear if any of Rupp's results were real or if they all were fakes. One of the people who commented was the British spectroscopist Robert d'Escourt Atkinson. He simply did not believe Rupp's results and gave his reasons. Rupp then claimed to have answered these objections. There is no evidence that Einstein ever realized that Rupp was a fraud. He simply, in a good natured way, tried to correct mistakes in Rupp's analysis but therefore adding credibility to Rupp's work. How long this might have gone on is not clear. But Rupp went one step too far.

In 1931, the British physicist, P. A. M. Dirac proposed the existence of the anti-electron — the "positron." The next year it was detected in cosmic rays. After that discovery, Rupp claimed that he was using positrons in his apparatus. This was such transparent nonsense considering his apparatus, that the jig was up. And indeed, his earlier experiment claims were examined and found fraudulent. This led to the letter I quoted in the beginning and to his retraction of his previous papers. After 1935, he never worked in physics again. Indeed, he had a nervous breakdown and entered a sanatorium. He died in 1979. One thing one can say in his favor, is that while getting his degree from Philipp Lenard and working in his institute, he never shared Lenard's virulent anti-Semitism. Lenard, a Nobel Prize winner who was a very early member of the Nazi Party, claimed that relativity was fake Jewish Physics, promulgated by the Jew Einstein. Clearly Rupp did not share these views. I have never found any comment by Einstein on this strange episode.

# 8 Pontecorvo

First Published Wall Street Journal, Feb.6, 2015

Appeared as a book review of Half-Life by Frank Close

Until the summer of 1950, when he was 37, the Italian-born physicist Bruno Pontecorvo seemed to be leading a charmed life. He had just accepted a professorship at the University of Liverpool. He was married to a beautiful Swedish woman. He had intelligent and active children. He was athletic — a first-rate tennis player, he once aspired to be the Italian champion — handsome and charming.

That summer he and his family left for a vacation on the Continent. They had planned to join his parents in Chamonix in France and then visit hers in Stockholm. But the day after they arrived there, they left again — this time for Helsinki — and disappeared. It was suspected that they were in the Soviet Union, which was only confirmed at a press conference that Pontecorvo held five years later.

It is a remarkable story — part physics and part Cold War intrigue — and it is wonderfully told in "Half-Life," a biography by the Oxford physicist Frank Close. Mr. Close works in the same fields that Pontecorvo did — elementary particles and nuclear physics — and is a prolific author. From his acknowledgments, it would appear that he has interviewed well over 100 people, including eight members of the

Pontecorvo family, among them the children who were with Pontecorvo when he went east. There is much about this tale that has the flavor of a le Carré novel, with the additional advantage that it is all true.

Pontecorvo, one of eight children, was born on Aug. 22, 1913, into a wealthy Jewish family in Pisa. They were in textiles. Nonetheless, his cousin Emilio Sereni, who after the war became Italy's minister of public works, was an active and committed communist. Pontecorvo first studied in Pisa to be an engineer but then switched to physics. In 1931 he switched his studies to Rome, where two years later Enrico Fermi accepted him as one of his small, elite group of physicists who were devoting themselves to experiments involving the neutron, which had been discovered in 1932 by the British physicist James Chadwick. Each member of the group had a nickname. Fermi was of course called the Pope, Franco Rasetti was called the Cardinal Vicar, and Emilio Segrè was called Basilisk, after a snake so venomous that it could kill at a glance. As the youngest, Pontecorvo was known as Puppy.

Not long after joining the group, Pontecorvo made an accidental discovery that in the hands of Fermi transformed nuclear physics and engineering. Having no electric charge, the neutron was the ideal probe for exploring the structures of different nuclei. Fermi and his group tried bombarding one element after another with neutrons, which sometimes caused the samples to become radioactive. Pontecorvo noticed that the radioactivity induced seemed to depend on the composition of the table on which the target sat. Wood accelerated the reaction, while marble did not. This had no explanation until Fermi repeated the experiment with paraffin. He realized that the collisions with positively charged protons in the paraffin and the wood had slowed the neutrons and that for quantum-mechanical reasons this enhanced the reactions. The discovery became the basis of the "moderators" used in nuclear reactors, which allow the fission process to proceed.

In 1936, Pontecorvo joined the laboratory of Irène and Frédéric Joliot Curie in Paris, where he met his wife, Marianne. While his life in

Fermi's laboratory had been politically neutral, Joliot was a committed communist and made no secret of it. Pontecorvo in turn became committed to the party. In 1940 he fled France, partly by bicycle, and he and his family made it to the United States. He got a job with an oil-drilling company in Oklahoma, where he invented a method, still in use, of using neutrons to detect promising rock formations. But in 1943, after having been branded an "enemy alien" for being Italian, he went to Canada to work on the design of a new reactor that used "heavy water" as a moderator. This type of reactor is especially well-suited to the production of plutonium. In 1949, he went to England and got a job at the U.K.'s atomic-energy lab in Harwell, where work on both the atomic bomb and a reactor was being carried out. Pontecorvo worked on the latter, and it was here that his life began to unravel.

On Nov. 9, 1949, Emilio Segrè — Basilisk — now at the University of California, went to an official at the Atomic Energy Commission and told him about the communist tendencies of members of the Pontecorvo family, for fear, Mr. Close writes, "that his former association with Pontecorvo could call his own loyalty into question." This information was given to the FBI, who informed the British. This led to concern but no action. But in January 1950, Klaus Fuchs, who was a colleague of Pontecorvo's at Harwell, confessed to being a Russian spy. He had been at Los Alamos and turned over to the Russians the detailed plans for the plutonium bomb that flattened Nagasaki. They used these planes to build their first device, which they tested in 1949. (Mr. Close thinks that the American Ted Hall was an even more important spy. I strongly disagree. I am familiar with what each of them turned over and there was no comparison.) Pontecorvo must have felt that the noose was tightening, and he bolted.

His life in Russia was made very comfortable. He had his work, which is what mattered. His wife had nothing and proceeded to collapse psychologically. Pontecorvo found himself a girlfriend. The work he did was what had made him famous among physicists. It involved the ghostly particle known as the neutrino. By the late 1950s it was realized that there were two different kinds — something that Pontecorvo had suggested. Neutrinos were thought to be massless. But Pontecorvo's great realization was that if neutrinos in fact had mass, then they could oscillate back and forth between the two types. This insight explained an anomaly in the number of neutrinos emanating from the sun that had long bothered physicists. I think that this might well have eventually won Pontecorvo a Nobel Prize, but he died, still in Russia, in 1993 before such a prize was awarded. Until nearly the very end, Pontecorvo never complained about the choice he had made. He always insisted that he had never been a spy, and Mr. Close could not find any definitive evidence that he was. My guess is that this is right. A year before his death, referring presumably to his choice to flee, he said to a journalist from the London Independent, "I was a cretin."

# III Science

# Three for the Road 9

First published in Inference (Volume 3, Issue 3)
www.inference-review.com

Max Born published a brief paper entitled "*Zur Quantenmechanik der Stoßvorgänge*" (On the Quantum Mechanics of Collisions)" in 1926.[1] This was one of the seminal papers in the history of the quantum theory. That same year, Erwin Schrödinger published the first of his papers on wave mechanics. It was not clear how to understand the wave function. A natural interpretation, at first endorsed by Schrödinger, was to think of waves as *Führungfelder* (guide fields) for particles, but this interpretation rapidly fell out of favor when it seemed to predict that the electron in the hydrogen atom could be guided to the moon. Born had studied the collisions between an electron and an atom. He suggested interpreting the wave function as the probability that the electron was scattered in some particular direction. But then Born corrected himself. It is the square of the wave function that is related to the probability.

This was an interpretation that certainly did not seem obvious on its face. At the 1927 Solvay Conference, Albert Einstein raised an issue, one recounted by Born:

*A radioactive sample emits α-particles in all directions; these are made visible by the method of the Wilson cloud chamber.*

*Now, if one associates a spherical wave with each emission process, how can one understand that the track of each α-particle appears as a (very nearly) straight line? In other words: how can the corpuscular character of the phenomenon be reconciled here with the representation by waves?*[2]

Born responded in terms a modern quantum theorist would find familiar:

*As soon as such ionization is shown by the appearance of cloud droplets, in order to describe what happens afterwards one must reduce the wave packet in the immediate vicinity of the drops. One thus obtains a wave packet in the form of a ray, which corresponds to the corpuscular character of the phenomenon.*[3]

This left open two questions. Could this collapse be described within quantum theory itself? And why should the path be more or less a straight line?

In 1929, Nevill Mott published a paper entitled, "The Wave Mechanics of α-Ray Tracks."[4] Take the inelastic scattering of an α-particle with a hydrogen atom. The particle's wavelength is $10^{-15}$ meters; the hydrogen atom is about $10^{-10}$ meters across. The final α momentum is **k**. We neglect the excitation energy of the hydrogen atom, so the scattering is elastic. The amplitude in the Born approximation is proportional to $\int \psi^*_{final} V \psi_{initial}$, where V is the Coulomb potential. The initial wave function is proportional to $e^{ikR}/R$, where R is the position of the α-particle. The final wave function is proportional to an outgoing plane wave. But if the vector that describes the location of the hydrogen atom in question is called **a** then the Coulomb potential is only large when R~**a**. If we make the various substitutions in the transition amplitude, we find an integral which has a factor $e^{ikR(1-\mathbf{ka})}$. Here **k** and **a** are unit vectors, and **ka** is their scalar, or dot, product. If **ka** = 1, all is well; if not, extreme oscillations appear over the electron orbital, effectively making the scattering amplitude zero. The final α must be in the direction of the initial one. We are in the process of generating a straight line. Mott shows this for two successive scatterings; no doubt one could continue this indefinitely.

But does this answer the question? Not exactly.

It surely will have struck the reader that there is no wave function collapse here. This is not surprising. Schrödinger's equation can never produce a collapsed wave function. Time evolution is a unitary operator and projections are not unitary. The ionization of some hydrogen atoms by collision with α does not produce an observable track. Bubbles are needed, condensates around the ions produced by the α-particle collisions. This process, which accounts for the cloud chamber tracks, is in practice irreversible; it very likely cannot be described using ordinary quantum mechanics.[5]

The Mott analysis was necessary, but not sufficient.[6]

Schrödinger had become skeptical of the foundations of quantum mechanics by 1935. If Paul Dirac had had any such qualms, he wisely kept them to himself. Schrödinger shared his feelings with Einstein who, in this respect, was a kindred spirit.

A particle's spin represents its intrinsic angular momentum. Aligned with a given coordinate axis, it can be either up or down: ↑ or ↓. Two electrons might be in a configuration of ↑↑, which has a total spin of one. But they also might be in a configuration of ↑↓−↓↑. In this configuration, the electrons do not have a definite spin, and we say, using Schrödinger's language, that the spins have become entangled. So long as we do not interfere with this state by performing a measurement, the entanglement persists, no matter how far the electrons are separated in space. But this has a consequence that Einstein referred to as a spooky action at a distance.

John Bell suggested another example to me. Consider identical twins separated at birth. When, later in life, they are reunited, it is discovered that they have very similar psychological profiles. Why? Because they have the same genes. If you believe quantum theory, however, there is no analogous explanation for the persistence of entanglement.

Take the two electrons in the configuration ↑↓−↓↑, and let them fly off in opposite directions. On the moon, an observer measures the spin of the arriving electron, and on the Earth, another observer does the same thing. There is no communication between the observers; only later do they compare notes. When they do, they find a remarkable correlation. Whenever the first observer has found ↑, the second observer finds ↓, and vice versa. If they have chosen axes separated by an angle θ, then quantum mechanics predicts a correlation that is proportional to −cos(θ). If, during the separation, one of the observers chooses to change the angle, the correlation will reflect this. No signal has passed between the two observers.

The spin-up and spin-down probabilities are, respectively, $\cos(\theta/2)^2$ and $\sin(\theta/2)^2$, which, of course, sum to 1. The anticorrelation of the two spins is $\sin(\theta/2)^2 - \cos(\theta/2)^2 = -\cos(\theta)$. Note that, when the angle is zero, the spins are perfectly anticorrelated, and when it is ninety degrees there is no correlation. No local hidden variable theory can reproduce these results for all angles. It is quantum mechanics *pure et dure*.

This argument can be reproduced in a simple form. I have at my disposal what I call Einstein robots. These can be programmed to do anything, except to communicate with each other at speeds greater than the speed of light. Can I program them to reproduce the quantum correlations? Imagine I have a bucket of spin ½ particles in singlet states — all electron spins are paired. I instruct the robots to take these particles and to bring them in pairs to two separate detectors. I have their magnets lined up; this setup will allow the robots to reproduce the anticorrelation of quantum mechanics. Then I rotate one of the magnets through θ. The robots have been programmed to reproduce the quantum correlation −cos(θ) when one of them encounters such a rotation. But now I play a trick on the robots. I rotate one detector by θ and the other by −θ. The robots cannot communicate, so each produces a correlation. When combined, the two give the correlation −2cos(θ).

The correct quantum correlation is −cos(2θ).

No signal can travel faster than the speed of light; one cannot claim a causal connection in the usual sense. One can note that this result is to be expected because of entanglement, but it cannot be explained in the same way as for the identical twins. The spins were entangled when the state was created and remained so until measurement. This is what Einstein and Schrödinger found impossible to swallow. Einstein spoke of spooky action at a distance, but this language is a relic of classical physics. You cannot explain quantum mechanics in terms of classical physics, although the converse is true.

Electrons are just one example of a particle with spin. All elementary particles have spin. This includes the photon, which has spin 1. As the photon propagates, its associated electric and magnetic field circulates around the direction of propagation. This is known as a state of circular polarization. The same trick done for pairs of electrons can be done for pairs of photons. The result is again an entangled state. This was shown by a remarkable experiment performed at the University of Science and Technology of China in Shanghai. At an altitude of five hundred kilometers, there is a Chinese satellite with a light-altering crystal on board. A laser signal was sent from the earth and produced pairs of polarized photons. These were transmitted to widely separated stations at high altitudes in Tibet. Their polarization remained correlated, just as quantum mechanics predicts. There is no explanation that would have satisfied Einstein. It is just quantum mechanics.

At the turn of the twentieth century, Max Planck introduced a set of natural units based on the fundamental constants of physics. Given in meters, the Planck length is

$$l_P = \sqrt{\hbar\, G/c^3} = 1.616 \times 10^{-35}$$

where $G$ is the gravitational constant. Planck's work was done before the advent of special relativity, so no one asked how this length transforms. But since fundamental constants are Lorentz invariant, it doesn't. How then can it be a length?[7] I once posed this question as a riddle. Only Freeman Dyson answered that this length is not measurable.[8] Suppose that there are two objects floating in space. Can they maintain their Planck distance long enough for it to be measured? Suppose the variation in the distance between them is of order $\delta$, and suppose that a measurement of the distance takes a time T. The uncertainty in the momentum is of order $\delta M/T$, so $M\delta^2 \geq \hbar T$. Let $\delta$ equal the Planck length $(G\hbar/c^3)^{3/2}$. If the distance between the two objects is D, then T must be greater than $D/c$ in order for the two objects to communicate. It follows that

$$D \leq GM/c^2.$$

The right-hand term should be familiar. It is the Schwarzschild radius of a black hole. The distance required to make the measurement is inside the black hole and thus unmeasurable. A similar sort of argument can be made for the Planck time. These units are not measurable. They are just numbers.

1. Max Born, "Zur Quantenmechanik der Stoßvorgänge" (On the Quantum Mechanics of Collisions), *Zeitschrift für Physik* 37 no. 12 (1926): 863–67. For a very nice discussion of this early history, see Rodolfo Figari and Alessandro Teta, "Emergence of Classical Trajectories in Quantum Systems: The Cloud Chamber Problem in the Analysis of Mott (1929)," arXiv: arXiv:1209.2665 (2012).
2. Quoted in Rodolfo Figari and Alessandro Teta, "Emergence of Classical Trajectories in Quantum Systems: The Cloud Chamber Problem in the Analysis of Mott (1929)," arXiv: arXiv:1209.2665 (2012): 6.
3. Quoted in Rodolfo Figari and Alessandro Teta, "Emergence of Classical Trajectories in Quantum Systems: The Cloud Chamber

Problem in the Analysis of Mott (1929)," arXiv: arXiv:1209.2665 (2012): 6.

4. Nevill Mott, "The Wave Mechanics of α-Ray Tracks," *Proceedings of the Royal Society A: Mathematical, Physical and Engineering Sciences* 126 (1929): 79–84. The paper is unnecessarily complex for the point being made here. Instead I will use Steve Adler's treatment — and notation — which makes the matter more transparent. I thank Steve Adler for sending me his paper and for discussions. The paper can be found online under the title "Mott Simplified" in the talks and memos section.

5. For a review, see Angelo Bassi and Giancarlo Ghirardi, "Dynamical Reduction Models," *Physics Reports* 379 (2003): 257.

6. As far as I know, there is no relativistic covariant theory of quantum mechanical measurements. This may be the real measurement problem.

7. The same question can be posed for the other Planck units.

8. Dyson discussed this in the context of the LIGO experiment but his argument does not make use of this. For details, see Freeman Dyson, "Is a Graviton Detectable?" *International Journal of Modern Physics A* 28, no. 25 (2013), doi: 10.1142/S0217751X1330041X.

# Bode's Law and the Trappists

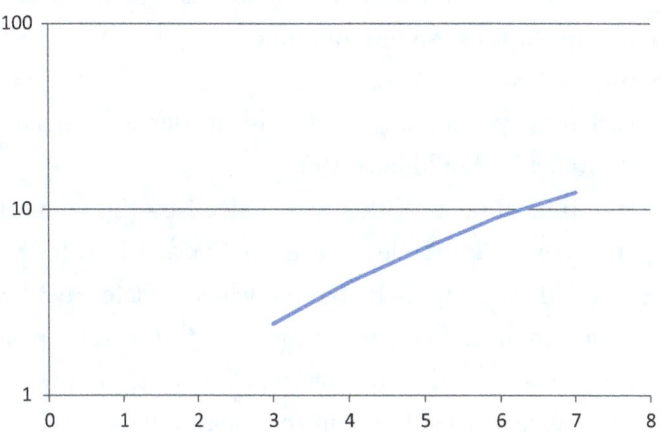

There are two things wrong with the title of this essay. "Bode's law" is not a law and it was not discovered by Bode. It is a "rule" sometimes obeyed by planets and sometimes not. It was first stated by the 18th century astronomer and scientific polymath, Johann Daniel Titius. In fact, as far as I can see, it is the only thing that bears his name in science (Titius, or Tietz, to give him his German birth name). It stems from a German translation he made of Charles Bonnet's *Contemplation de la Nature* to which he added

> Take notice of the distances of the planets from one another, and recognize that almost all are separated from one another in a proportion which matches their bodily magnitudes. Divide the distance from the Sun to Saturn into 100 parts; then Mercury is separated by four

such parts from the Sun, Venus by 4 + 3 = 7 such parts, the Earth by 4 + 6 = 10, Mars by 4 + 12 = 16. But notice that from Mars to Jupiter there comes a deviation from this so exact progression. From Mars there follows a space of 4 + 24 = 28 such parts, but so far no planet was sighted there. But should the Lord Architect have left that space empty? Not at all. Let us therefore assume that this space without doubt belongs to the still undiscovered satellites of Mars, let us also add that perhaps Jupiter still has around itself some smaller ones which have not been sighted yet by any telescope. Next to this for us still unexplored space there rises Jupiter's sphere of influence at 4 + 48 = 52 parts; and that of Saturn at 4 + 96 = 100 parts.

Apart from the missing planet the reader will notice that the next planet predicted should have 196 "parts."

This observation of Titius was noticed by the young German astronomer Johann Elert Bode, who at first added it to his own book on astronomy without proper attribution, which he later rectified. Bode publicized this result and urged a search for the missing planet. The asteroid which was named Ceres and which was at the predicted distance was discovered in 1801, but in the meanwhile an even more significant planetary discovery had been made at just where the rule said it should be. It was discovered in 1781, by the British astronomer William Herschel. He wanted to call it "George" after the king, but it was in the event given the name Uranus. Herschel had made no use of the rule in his discovery, but it gave the rule an apparent legitimacy, that it is until Neptune and Pluto were discovered which didn't really fit. Nonetheless veritable forests have been cut down to supply the paper on which "explanations" of the rule have been printed. Most traditional astronomical journals stopped accepting papers on the subject and then came the exoplanets.

By now, a few thousand planets have been observed orbiting their suns. It is natural to ask if they obey anything like a Titius-Bode rule. The rule has now been formulated in algebraic form. If $a$ is the distance of the planet from the sun and $b$ and $c$ are parameters, then the rule would read

$$a_n = b(c)^n$$

where n is the nth planet. Remembering from high school note that

$$\log(a_n) = \log(b) + n\log(c).$$

Thus

$$\log(a_{n+1}) - \log(a_n) = (n + 1 - n)\log(c) = \log(c),$$

or in words, if this rule was obeyed, then this difference must be the same for all adjacent planets. This is one of the criteria that was used by the Australian-American Charles Lineweaver and his student Timothy Bovaird to analyze the orbits of the exoplanets. They find that in many

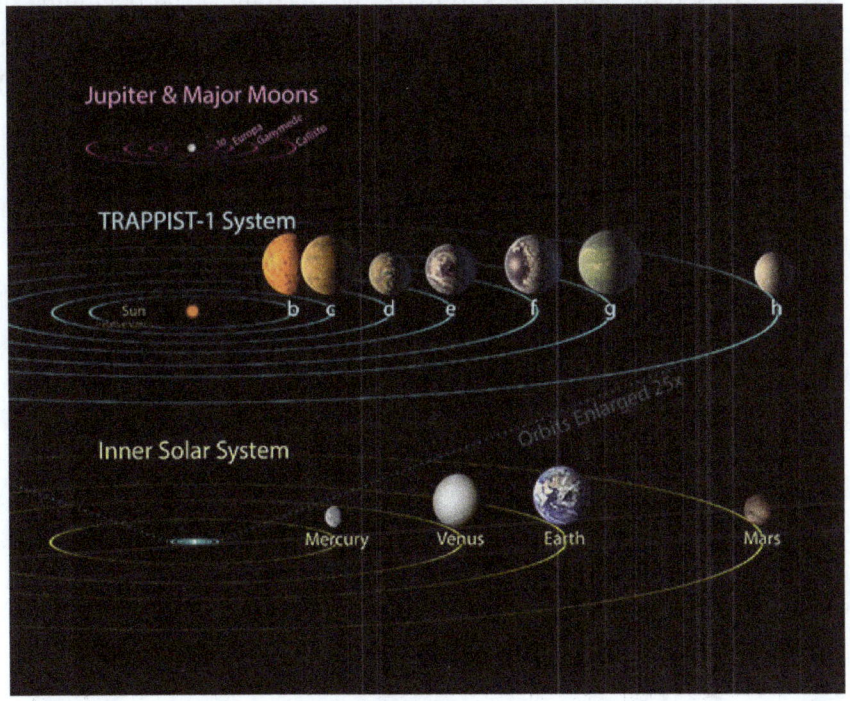

Credits: NASA/JPL CalTech / R. Hurt. T. Pyle JPAC

cases the Titius-Bode rule holds as well or better than it does for the solar system. This appears to be true of the seven planets just found in the galaxy Aquarius, circulating around their dwarf star, always with one face locked onto their sun. In fact, it is a remarkable fit, even better that our own Solar system.

Here are the semi-major axes observed for the seven planets. The reader can entertain him or herself by taking the log differences as seeing how well the Titius-Bode law is obeyed.

- o  TRAPPIST-1 b, planet, semi-major axis: 0.0111±0.0003 AU
- o  TRAPPIST-1 c, planet, semi-major axis: 0.0152±0.0005 AU
- o  TRAPPIST-1 d, planet, semi-major axis: 0.0214+0.0007−0.0006 AU
- o  TRAPPIST-1 e, planet, semi-major axis: 0.0282+0.0008−0.0009 AU
- o  TRAPPIST-1 f, planet, semi-major axis: 0.0371±0.0011 AU
- o  TRAPPIST-1 g, planet, semi-major axis: 0.0451±0.0014 AU
- o  TRAPPIST-1 h, planet, semi-major axis: 0.063+0.027−0.013 AU

I used these numbers to make the diagram at the beginning of this essay. The reader will note that the vertical axis is logarithmic. If this form of the Titius-Bode law is true, this plot should be a straight line. I leave it to the reader to decide. If it is true one wonders why.

# 11

## Advanced Quantum Mechanics

*"I deduce two general conclusions from these thought-experiments. First, statements about the past cannot in general be made in quantum-mechanical language.*

*We can describe a uranium nucleus by a wave-function including an outgoing alpha-particle wave which determines the probability that the nucleus will decay tomorrow. But we cannot describe by means of a wave-function the statement,*

*"This nucleus decayed yesterday at 9 a.m. Greenwich time".*

*As a general rule, knowledge about the past can only be expressed in classical terms. My second general conclusion is that the "role of the observer" in quantum mechanics is solely to make the distinction between past and future. The role of the observer is not to cause an abrupt "reduction of the wave-packet", with the state of the system jumping discontinuously at the instant when it is observed.*

*This picture of the observer interrupting the course of natural events is unnecessary and misleading. What really happens is that the quantum-mechanical description of an event ceases to be meaningful as the observer changes the point of reference from before the event to after it.*

*We do not need a human observer to make quantum mechanics work. All we need is a point of reference, to separate past from future, to separate what has happened from what may happen, to separate facts from probabilities"*

*Freeman Dyson*

In 1951, I entered graduate school at Harvard in mathematics which had been my undergraduate major. But in truth, my real interest was physics and even the mathematics I studied, such as Hilbert space, had an orientation in physics. Indeed, after I took my master's degree, the mathematics department chairman told me that I had to choose between mathematics and physics. I chose physics. By this time, I had taken some courses in the physics department, including a sort of introduction to quantum mechanics offered by Julian Schwinger, one of the giants in the revolution in quantum electrodynamics. I say "sort of" because there was a sort of Jekyll and Hyde development of the course. The Doctor Jekyll part of the course was the fall semester, where Schwinger gave a marvelous introduction to the experiments that led to the quantum theory and some of the controversy. I have an ineluctable memory of his description, of how Einstein had given an argument that purported to show that the uncertainty principle connecting energy and time could be violated. Bohr pointed that Einstein had overlooked the effect of gravitation on time, his own discovery, and when this was taken into account, quantum mechanics was saved. Then came the Mister Hyde part.

It turned out that Schwinger was developing his own version of the quantum theory. It involved using propagators — Green's functions — to describe the time evolution of matrix elements. It could not have been a worse introduction to the theory. The Schrödinger equation was somehow derived and then ignored. In one lecture, Schwinger got stuck and had to defer to the next lecture. It got so bad that some of us went to MIT to audit Viki Weisskopf's lectures. I will never forget the first one. Viki walked in and said, "Boys" — there were only men — "I had a wonderful night."

There were raucous cheers. "Yeah Viki!"

"No", he said, "It is not what you think. I have finally understood the Born approximation." When I think of all the things that Schwinger could have taught us — not the least of which was the Dirac equation and some of its applications — I grieve at the loss. But as it happened, some of the void was filled by an unexpected source — Freeman Dyson.

Dyson had come from Britain to do his graduate work with Hans Bethe at Cornell. Bethe had appreciated his extraordinary abilities and Dyson had returned from an interim in England to teach at Cornell. Indeed, he was teaching advanced quantum mechanics. His lecture notes were circulating in a sort of samizdat form. In think I paid two dollars for mine. I am sure that I have them around somewhere. But I don't need to search because an elegant edition of them has been published by World Scientific. Various errors and infelicities in the original notes have been corrected by David Derbes. There is an appendix describing these corrections if the reader is interested in seeing what was in the original.

Let me begin by saying what this book is not. There are very few problems to be solved and no discussion of what people sometimes refer to as the interpretation of quantum mechanics. In this respect, it resembles Dirac's classic Quantum Mechanics, although Dirac proposes no problems. When once asked about it, Dirac said he thought that it was a good book, but it lacked a first chapter — the interpretation of the theory. If you want both things, I suggest Steven Weinberg's Lectures on Quantum Mechanics. The reader may become disconcerted to learn that Weinberg has not found a satisfactory interpretation of quantum mechanics. I agree with him and I think Dyson would agree if you insist on using the quantum theory to describe the past. The past in the quantum theory has always been vexed. You can easily construct examples which apparently violate the uncertainty principles. For example, suppose you have a system that ejects a particle at a precise location, from which it travels unimpeded to a detector which measures its momentum precisely. If you argue that that was its momentum when ejected, you get a violation of the uncertainty principle. The Bohrean quantum mechanist would say if you haven't measured it, it does not exist. The discussion of this has filled volumes. In his book, Dyson says nothing about any of this.

The book begins by making the case for field theories, which are of necessity, many-particle theories. Consider a spin zero particle that obeys the Klein Gordon equation ψ

$$\frac{1}{c^2}\frac{\partial^2 \psi}{\partial t^2} = \frac{\partial^2 \psi}{\partial x^2} - \mu^2 \psi.$$

This is a second order equation, which means that both the wave function and its time derivative can have either sign. This means that the probability density derived from the conserved current

$$\rho = \frac{ih}{2mc^2}\left(\psi^* \frac{\partial}{\partial t}\psi - \psi \frac{\partial}{\partial t}\psi^*\right)$$

is not necessarily positive. It must be re-interpreted as a charge density in a many body theory, i.e., a quantum theory of fields.

Dyson gives a brief but very clear introduction to the Dirac equation and how it is solved in the case of the hydrogen atom — the classic case. He notes that with the Dirac equation, the problem was that it had negative energy solutions. This was considered so serious, that in a famous review article, Wolfgang Pauli stated that the equation was unphysical. Dyson gives a simple argument against negative energy particles. They can never be stopped by ordinary matter at rest. They keep gaining energy. The solution to this dilemma is well-known. A negative energy electron is really a positively charged electron — the positron in disguise. The positron was first observed by Carl Anderson in 1932. Dirac was awarded the Nobel Prize in 1933. I do not think that the dates are a coincidence.

Most of Dyson's book has to do with quantum electrodynamics. Given the date of the original lectures, 1951, this is hardly a coincidence. While there was other very important physics going on, it was the results of quantum electrodynamics that made the deepest impression — at least at Harvard, where Schwinger, one of its developers, was resident. Indeed, there was a cigarette smoke-filled room in the basement of the physics building, where many of the Schwinger's students did their work calculating various twists and turns that Schwinger had not bothered to calculate. There was a calculation done by others that did not agree with experiment and Schwinger thought it might be wrong. He put a student Charlie Sommerfield onto the subject and he found the mistake, which made him for at least a short while a local celebrity. I was not in the basement. I was actually working with a young instructor named Abraham Klein on the theory of mesons. None of these meson calculations made much sense because of the magnitude of the constant that characterized the couplings of mesons to nucleons — about 10 — which ruled out perturbation theory although people did it anyway. On the other hand, the fine structure constant $\alpha = e^2/\hbar c \sim 1/137$, which characterized electrodynamics, was small enough to encourage an expansion in it. Indeed for a while, Dyson thought that this expansion might be convergent, meaning that various functions of $\alpha$ would be analytic so they could be extended to negative $\alpha$. This would be a real disaster since electrons would attract electrons and positrons would attract positrons and the bottom would fall out.[1] The perturbation series cannot converge. This was not mentioned in Dyson's notes.

While I was not working in anything related to quantum electrodynamics — my thesis had to do with the deuteron — nonetheless I felt that I should learn about it. I might note that when I told a colleague I was going to call my thesis Deuteronomy, he said that he was calling his Exodus. The obvious places for me to look were Schwinger's papers. I have some of them in front of me as I write this. They looked and still look impenetrable. Oppenheimer once said of Schwinger, who had been with him briefly as a postdoctoral, that while most people write papers to show how to solve a problem, when Schwinger writes a paper it is to show you that only he can solve the problem.

I got some friendly advice from another student — read Feynman — which I tried to do. The mathematics of his papers was less formidable but I could not make head nor tail of what he was doing. There were a few odd diagrams that were somehow connected to the equations. Many years later I witnessed high in the Himalayas a religious ceremony performed by the local monks. I watched with total incomprehension, just the feeling I had when I tried to read Feynman's papers. I was about to give up on the subject when I came across Dyson's 1949 paper, "The Radiation Theories of Tomanaga, Schwinger and Feynman." When I read this paper, much of the subject fell into place. The diagrams began to make sense and I could see how they were connected to the underlying theory. It should not be surprising that Dyson's lectures delivered a couple of years later are an adumbration of his papers — there was more than one — amplified for the benefit of his students. In fact, one might well call his lectures and the book an introduction to quantum electrodynamics (QED) because that is what it is.

Quantum electrodynamics is nearly as old as the quantum theory itself. The problem that it had was fully revealed in the 1930's, namely when you tried to calculate something beyond the lowest order, the answer was generally infinite — that is meaningless. To take an important example. When you do a calculation in QED you insert a parameter that represents the electron mass $m_0$. But if you were to measure the actual mass it would not be this bare mass, but a mass m that has been influenced by the sea of electrodynamic fluctuations in which the electron moves. If you try to calculate the relationship between m and $m_0$, you find that m differs from $m_0$ by an infinite amount, which is an absurdity. The same observation can be made about the electron's charge. While this problem was well-known in the 1930's, there was no incentive from experiments to do much about it, except complain. That all changed after the war, when physicists using techniques, that had been learned from working on things like radar, began doing measurements that could only be explained by taking into account these corrections to the lowest order in $\alpha$. Before the war, the Dutch theoretical physicist Hans Kramers had given an important clue as to how these infinite corrections might be handled — mass renormalization.

Suppose you did a calculation using the bare mass that resulted in a term that looked like a kinetic energy. But if you had written the actual kinetic energy in terms of the bare mass plus this correction, you have to subtract off this infinite term to extract the finite physical answer, since you have already taken it into account. As a result of Dyson's work, it was known that in QED, the only other quantity that had to be renormalized was the electric charge. A theory which can be rendered finite by a finite number of such renormalizations is called "renormalizable." The theory of gravitation is notoriously not renormalizable. We lack experimental guidance as to how to proceed, so speculation runs rampant.

Before starting the heavy lifting of presenting many of the standard QED calculations, such as the Lamb shift and the magnetic moment of the electron, the scattering of light by light is only glossed over briefly in an appendix. Dyson makes the following observation,

"In this course we follow the pedestrian route of logical development, starting from general principles of quantization applied to covariant field equations and deriving from these principles first the existence of particles and later the results of the Feynman theory. Feynman by the use imagination and intuition was able to build a correct theory and get the right answers to problems much quicker than we can. It is safer and better for us to use the Feynman space-time pictures not as the basis for our calculations but only as a help in visualizing the formulae which we derive rigorously from the field theory. In this way we have the advantages of the theory, its concreteness and its simplification without the logical disadvantages."

This is just what I was missing when I tried to read Feynman's papers all those years ago. Where did his formulae and space-time pictures come from?

Where then are we left? Dyson's book is a wonderful introduction to QED as it was in the early 1950's. What held then still holds now but many things have moved on. As for QED itself, there is the work that was began by Murray Gell-Mann and Francis Low on what is totally misnamed the "renormalization group." The basic physical idea can be illustrated by the electron charge. We might try to measure it by scattering one electron from another. If we did, we would discover that the charge we measure depends on the energies of the electrons. This results from what is called "vacuum polarization." While the vacuum is the state of lowest energy, it is not empty. It swarms with electrons and positrons that have a brief appearance consistent with Heisenberg's uncertainty principle. The positrons shield the charge and this means that what we measure depends on the energies, since energetic electrons come closer to each other. Things are observed at a different length scale and the effective charge reflects this. Any more modern text on QED would have this. But there is something even more fundamental.

At the time of Dyson's lectures, QED was a little island unto itself. But it was realized that it did not live in isolation. It had to be combined with the theory of weak interactions of the kind that produce beta decay to make a consistent renormalizable whole. This was first worked out by Steve Weinberg. Since weak interactions are weak, they don't much influence the sort of effects that Dyson computes in his lectures. But the converse is not true. The theory of beta decay for example must include radiative corrections. Nonetheless a student even now would, in my view, profit from studying QED and for this purpose Dyson's lectures are superb.

In writing these recollections, I feel something like a sorcerer who has summoned up a djinn. In this case the djinn is Julian Schwinger and his quantum mechanics course. What has prompted this reflection is a new edition of Schwinger's text on the quantum theory to be brought out by Springer. The new edition is to commemorate the centennial year,1918, of Schwinger's birth. He was born on February 12 of 1918. I never saw Schwinger age. When I knew him, I was a student and post doc, he was in his thirties. The last time I remember seeing him was sometime after Robert Oppenheimer's death in 1967. His former student and later collaborator Robert Serber had organized a little conference of invited participants at the Institute for Advanced Study. I suppose it was to be some sort of reincarnation of the Shelter Island conference of 1947, where the new experimental results that led to the renaissance of QED were announced. The only thing that happened at our conference, that turned out to be memorable, was the presentation by John Wheeler of the no hair theorem for black holes, which shows that they are characterized by very few parameters. There was not much interest at the time, since black holes were not considered to be of much importance.

In attendance at this conference, were among others, Schwinger, Feynman and Dyson. I have an ineluctable memory of being shuttled in a bus back to our hotel in the town of Princeton. Feynman chose the occasion to deliver some sort of impromptu lecture. I recall Schwinger listening to this performance with a bemused look on his face. That was the last time I saw him. He died in 1994 at the age of 76. But I did hear about him and what I heard sort of depressed me. Some of this came from I. I. Rabi. Rabi had played a crucial role in Schwinger's career. It happened in 1935, when Rabi was trying to understand the famous paper by Einstein, Podolsky and Rosen which was meant to show that quantum mechanics could not be the complete description of nature. He had summoned one of his students Lloyd Motz to explain it to him and Motz said that there was this kid outside the office he would like to invite in. It was the sixteen-year old Schwinger who settled a point on which they were stuck. Rabi said to himself "Who the hell is his?" He discovered that Schwinger was at City College where he was doing badly. He had failed English. When Rabi brought this up, Schwinger said that he did not have time to do the themes. He was already publishing his first physics papers. Rabi brought him to Columbia where he did his thesis and became the theoretical adjunct to Rabi's experimental group.

Because of this history, Rabi became a kind of father figure to Schwinger. He told me that Schwinger in later years was quite unhappy because the younger generation was not paying much if any attention to his ideas. Perhaps this helps to explain what seems almost like obsessive behavior in Schwinger's last years. He worked on what is known as cold fusion. To make two light nuclei fuse and produce energy they must normally be put in an environment like the interior of a star where there is enough ambient energy available, so that these nuclei can surmount the Coulomb barrier which prevents their fusing. But in 1989, two researchers Martin Fleischmann and Stanley Pons claimed that they had performed a table top experiment that produced fusion at room temperature. As far as I can tell no one has been able to reproduce their experimental result. Nonetheless, Schwinger decided to take up the case. He tried to publish a paper called "Cold Fusion: A Hypothesis" In the Physical Review Letters. The referee reports were so nasty that Schwinger resigned from the American Physical Society. His final views were completely expressed in a lecture he was to give in 1994. He was too sick to give the lecture but one can find the text online. I have read the lecture and do not feel that I have the technical competence to evaluate it. The whole business saddens me and I prefer to remember the Schwinger that I knew when he was at the height of his powers.

The edition of Schwinger's *Quantum Mechanics: Symbolism of Atomic Measurements* is not the centenary edition, which is in preparation as I write this, but the original edition. Both editions are edited by Berthold-Georg Englert, a physicist who had the chance to work with Schwinger at UCLA. Schwinger went there from Harvard in 1972 to the surprise of many people, myself included. I thought he was a permanent fixture in Cambridge. What Englert has done, was to assemble the notes of the course or courses that Schwinger taught on quantum mechanics at UCLA. They are separated into the Fall Quarter and Winter Quarter. One of the first thing one observes about them is the plethora of student problems. I am told that the anniversary edition will have the solutions to these problems. The question that came immediately to my mind is, to what students was this course addressing? It is clearly not an introduction to quantum mechanics. It is an introduction to Schwinger's version of quantum mechanics and let the reader beware!

There is a substantial prologue which resembles slightly the Doctor Jekyll part of the course I took with Schwinger in the day. He was much more detailed than this prologue. Maybe because his course was meant to be an introduction to the subject. In the prologue, Schwinger lays special emphasis on the difference between quantum mechanical measurements and classical ones. He writes, "Physics is an experimental science. It is concerned only with those statements which in some sense can be reduced to experiment. The purpose of the theory is to provide a unification, a codification, or however you want to say it, of those results which can be tested by means of experiment. Therefore, what is fundamental to any theory of specific departure is the theory of measurement within that domain." What was disappointing to me was that there is no discussion in the book of the paradoxical fact that standard quantum mechanics does not and cannot describe the measurement of quantum mechanical properties. I will explain.

Let us suppose we have a system described by a wave function $\psi(x, t)$, which obeys the Schrödinger equation

$$i h \frac{\partial}{\partial t} \psi(x.t) = H \psi(x, t),$$

where H is the Hamiltonian. The solutions are perfectly time reversible. If we know $\psi$ at some time we can both predict and retrodict its value at any past or future time. Associated with the system are "observables" — momentum, angular momentum and the like. These are represented by

Hermitian operators, A, B, etc. These operators have eigenvalues a, b, etc. and these eigenvalues are the predicted results of any possible quantum mechanical measurement. The wave function ψ can be expanded in a basis of any of these eigenfunctions, i.e.,

$$\psi = \Sigma \, \psi a c_a,$$

where the $c_a$ are complex number coefficients. We have known since the work of Max Born that the absolute values of these coefficients represent the relative probability of finding the value a, in a measurement of the system. In consequence of this measurement the wave function, Ψa is projected out of the sum, something that is referred to often as the "collapse of the wave function". This term is never mentioned in Schwinger's book. This projection is irreversible and therefore cannot be described by the Schrödinger equation and hence it eludes standard quantum mechanics.

This was well-known to people like Bohr. He claimed that the measuring apparatus was classical while the state it was measuring was quantum mechanical. But he was never able to give a precise boundary between the two. Most working physicists simply ignore this issue since they have no trouble in making the distinction on a practical level. Schwinger also completely ignores the question, which I think is less excusable in a text that it supposed to deal with measurement among other things. Weinberg on the other hand does deal with the question but ends up by concluding that none of the proposals on offer is satisfactory. Let me begin with one that he does not consider, which is the quantum mechanics of David Bohm.

In 1952, when Bohm published the first of his papers, I had a very brief discussion with Schwinger about it. I said it looked too complicated. He said it looked too simple. This is also what Einstein said. What Bohm did was to replace the quantum mechanical time evolution by the evolution of a particle that obeys Newton's law. The theory is manifestly non-relativistic. The quantum mechanics comes in, because there is a "quantum potential" determined by the wave equation. This way, all the results of non-relativistic quantum mechanics are reproduced. There is nothing special about an apparatus. The particle simply interacts with it under the influence of the quantum potential. It was first clearly pointed out by John Bell, that if you had say two interacting particles the behavior of one was influenced by the instantaneous behavior of the other. This is a clear violation of relativity so it is difficult to see how the formulation could be relativized. Within its limited domain it, gives one another way of approaching non-relativistic quantum mechanics.

I once heard a lecture by Eugene Wigner on the subject. He concluded that one might attach a small piece to the Hamiltonian, that would produce the wave function collapse. Variations of this have been suggested. In his discussion Weinberg does not consider this possibility either. Rather he focuses on what is called the "many world" or "many history" interpretation. The former was introduced in a 1957 Princeton PHD thesis by a student of Wheeler named Hugh Everett. The basic idea is that in a measurement, the wave function does not collapse but one branch is singled out and the rest propagate freely in the world, only to split into more branches and still more, as time goes on. I do not see anything logically wrong with this but it certainly is inelegant. A variant has been studied by Murray Gell-Mann and Jim Hartle. They refer to it as the many history variant. It accomplishes the same thing. It takes advantage of the fact that between an initial and final state of a quantum system, there are many possible paths or histories. As observers we follow one. They have worked this out in some detail but I do not think it has been made relativistic. You can consult Weinberg's book for details. He concludes that none of these approaches is really satisfactory and that it may require a new theory, to which quantum mechanics is an excellent approximation.

So we have three books. Dyson's book is a superb introduction to quantum electrodynamics. He had written to me that he is surprised that the theory has stood up all these years. At the time of the notes he felt it would last about five. It has predicted effects to parts in a trillion. Schwinger's book is an exhibition of virtuoso mathematics being applied to familiar problems. Weinberg's book is the one to choose if you want to learn how quantum mechanics is used by the common man.

## Endnotes

1. There is a tricky point here. One must balance the kinetic and potential energies. If N is the number of particles, then the kinetic energy is proportional to N. But since the Coulomb force is long range the potential energy is proportional to N(N-1) and hence overwhelms the kinetic energy. That is what causes the bottom to drop out.

# Gian Carlo 12

The neutron was discovered in 1932 by James Chadwick. Fermi named it the "neutrone" — the big neutral one — to distinguish it from the "neutrino" — the little neutral one. As far as I know, Heisenberg was the first person to successfully apply quantum mechanics to the structure of nuclei which resulted in a series of papers beginning in 1932. Prior to that, Gamow and others had used quantum mechanics to explain alpha particle decay as an example of quantum tunneling. Gamow had at about the same time introduced a liquid drop model of the nucleus, but his nucleus had relativistic electrons in it and he could not make sense of the model. When Heisenberg came back to it, his nucleus had only neutrons and protons and things worked much better.

Heisenberg made some assumptions that were consistent with what was known at the time. There were two sources of force. There was a short range powerful nuclear force that acted only between neutrons and protons, not protons protons, or neutrons neutrons. There was a coulomb force that acted between the protons. He discussed different possible forms for the nuclear potential. One of these was an exponential of the form $e^{-\mu r}/r$ where $\mu$ sets the scale. It was not until 1935 when Yukawa proposed that the force was due to the exchange of mesons, because of the energy time uncertainty principle, the meson mass sets the scale. Wick was the first one to point this out. Heisenberg, at least in the paper I read, is interested in the ratio of the number of neutrons to the number of protons. In the absence of the repulsive Coulomb force between the protons, this ratio would be sensibly constant. But because of the Coulomb force, more neutrons need to be added to keep the nucleus from breaking up. He writes this coulomb correction — in modern notation — to be proportional to $Z^2/A^{1/3}$. This reflects the liquid drop model. The volume of the "drop" is proportional to the number of nucleons A. Thus the radius is proportional to $A^{1/3}$ hence the form. He finds the constants empirically that seem to fit the data. However, Gian-Carlo is interested in the binding energies. His paper *Sulle Propieta Della Materia Nucleare* was published in the Il Nuovo Cimento in 1934. The result it contains is so fundamental to the subject that it is almost never credited to Wick. An exception is the paper *Zur Theorie der Kernmassen* which was written in 1935, but published the next year by C. F. von Weizsäcker. It was this paper that introduced the semi-empirical mass formula. He gives Wick a footnote. The only textbook I could find that credits Wick is Blatt and Weisskopf's *Theoretical Nuclear Physics*. I now turn to this.

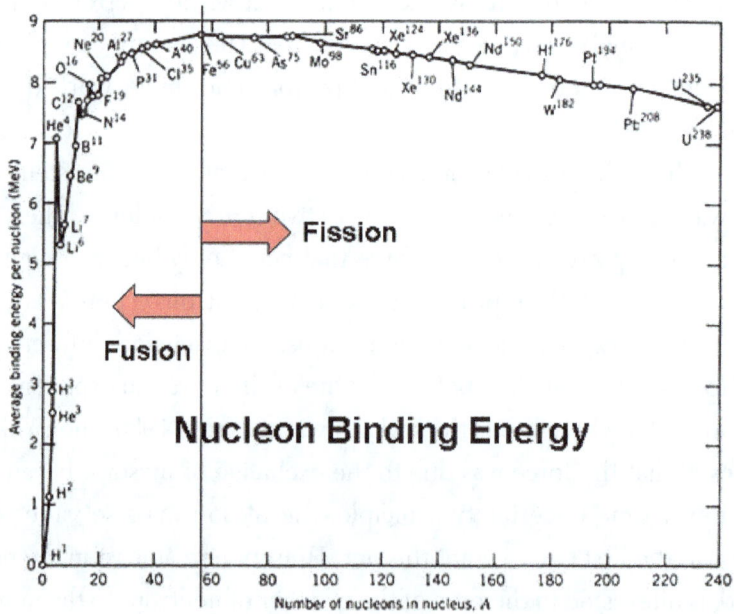

The figure above is a modern plot of the binding energy per nucleon — neutrons and protons — as a function of the total number A. I have chosen an example where fission is exhibited. I will explain this later. The remarkable thing is that no one noticed this until the end of 1938. It was staring people in the face. You can see that curve rises sharply until about A=56 — iron — and then falls off gently. This is what has to be understood. The liquid drop model in its simplest form — which in any event applies only to heavy nuclei — predicts that the binding energy per nucleon in this region should be a constant. One might argue as follows. In a real liquid drop, the heat of vaporization should be proportional to the number of molecules in the drop. Hence in our analogue model the binding energy should be proportional to the number of nucleons A. Hence B/A should be a constant. The first correction to this is the Coulomb term which reduces the binding energy. If we think of the nucleus as a uniformly charged sphere, then the Coulomb energy is proportional to $Z^2/r$, where $r$ is the radius of the nucleus. But we have argued that in the liquid drop model with constant density, the radius is proportional to $A^{1/3}$, hence this term is proportional to $Z^2/A^{1/3}$. Wick made an amendment to this which was extremely important. He noted that in the first term all the nucleons in the volume were being treated equally. But this is wrong. The nucleons on the surface do not have the same number of partners. Hence they are less effective in the binding. Hence one must add a surface term which is proportional to $A^{2/3}$, the surface area. Actually he assumes that the proton densities are constant so his radii are defined in terms of Z. Weizsäcker writes, using more modern notation as

$$B = -m_n A + C_1 A^{2/3} + C_2 Z^2/A^{1/3},$$

And this part of the mass formula has been written this way, with Gian-Carlo's amendment, ever since. By now, a number of quantum mechanical effects have been added and the formula has become increasingly more complex. One must use the data to find the best fit for the many constants. I doubt that many people could tell you who was responsible for the second term. At about the same time, Wick explained some new beta-decay data by noting that it must be a proton decaying into a positron, a neutron, and a neutrino. Which was an extension of Fermi's theory of the weak interactions. This too, is part of our given.

Let me now explain what everyone missed until the end of 1938 and then I will tell you about Gian-Carlo's trajectory. A glance at the curve of binding energy per nucleon, shows a gentle fall off of about 1 MeV between iron and uranium. Since there are some 200 nucleons involved, there is a disposable energy of about 200MeV if you could split the uranium nucleus. No one noticed this. Not Fermi, Not Heisenberg, Not Bohr—no one at least until the end of 1938. That year, the physicist Lise Meitner was in Sweden. She had Jewish ancestry and had been forced to leave Berlin where she had been working. Her nephew Otto Frisch had found refuge in Bohr's institute in Copenhagen. The two had decided to meet in Sweden during the Christmas vacation. While in Berlin, Meitner had collaborated with the radio chemist Otto Hahn. He was continuing some experiments he had started with Meitner and another radio chemist named Fritz Strassmann. These involved bombarding uranium with neutrons. They had found a result which they did not understand at all. They had produced the lighter element barium.

Hahn had written Meitner a letter asking for help. She had this letter when Frisch met her. That was all she wanted to discuss. Frisch tried to distract her by suggesting that maybe Hahn had made a mistake. Meitner said that Hahn did not make that kind of mistake. They went for a walk in the woods, Frisch on cross country skis and his aunt trotting along on foot. At some point, they had the realization of what I said could have been had, several years earlier. By adding the charges, they knew that along with the barium came krypton, an inert gas that simply left the scene. Weizsäcker had been a postdoctoral student with Meitner, so she knew the mass formula and its parameters so they could make a more precise estimate of the energy. Then Frisch went back to Copenhagen and told Bohr, who simply could not believe they had been so stupid as not to have seen this earlier. I do not much like counterfactual history but I shudder to imagine what it would have meant, if it had been discovered in say Germany some years earlier. It was there in the mass formula.

Gian-Carlo spent the war-time years in Italy. After Fermi left Rome in 1938 for the United States, he recommended Gian-Carlo to succeed him. The war-time years in Rome must not have been easy. I never heard him speak of them. He came to the United States in 1946, first to Notre Dame and then Berkeley. This was the McCarthy era and Gian-Carlo ran afoul of the loyalty oath that the trustees insisted, that all faculty members in the California state system must sign. He had had enough of that sort of thing in Fascist Italy and refused to sign. He lost his job and moved on for a while to the Carnegie Institute of Technology, before coming to Brookhaven in 1957. He spent a year — 1953 — at the Institute for Advanced Study and was the co-author of a famous paper with Eugene Wigner and Arthur Wightman, on the concept of intrinsic parity. He became a professor at Columbia in 1965, from which he retired in 1977. Until his death in 1992 he took up a position in the Scuola Normale in Pisa. After Brookhaven, I did see him from time to time and wish now I had spoken to him more. There is now a commemorative gold medal in his name, which honors theoretical physicists who have made outstanding contributions in the theory of elementary particles.

## Endnotes

1. One must distinguish between what I would refer to as "intrinsic" and "induced" couplings. The former are in the nature of the force itself and the latter are "induced" in the course of perturbative calculations. We considered only the former. An example of the latter is a negative muon emitting a virtual negative pion and a neutrino. The proton absorbs the pion becoming a neutron. The appears as a pseudo-scalar in the capture rate. There is now a vast literature on these effects and their experimental consequences. It is beyond the scope of this brief note to comment on it.

# IV  Nuclear Weapons

## An Error 13

> "The targets are too small"
> — J. Robert Oppenheimer

> "…but they cannot decide the outcome of a war, since atomic bombs are quite insufficient for that. Of course monopoly ownership of the secret of the atomic bomb creates a threat, but against it there exist at least two means: a) monopoly ownership of the atomic bomb cannot last for long; b) the use of the atomic bomb will be prohibited."
> — Joseph Stalin 1946.

After the atomic bombing of Hiroshima and Nagasaki in August of 1945, Robert Oppenheimer became the avatar of the new nuclear age. He was perfect for the part. He had the somewhat anguished look of a sadhu and had even studied Sanskrit. The things he said seemed profound, even though one was not quite sure of their meaning. He enjoyed the role. He had quelled a rebellion at Los Alamos, when after the German surrender, some of the physicists wanted to stop making the bomb. He felt one had to see it through, if for no other reason to see if it worked. President Truman decided he wanted to meet Oppenheimer. The meeting took place at 10:30 a.m. on October 24, 1945. Truman wanted Oppenheimer's help in getting the May-Johnson bill passed, which would have given the military the sole jurisdiction over atomic energy. It became clear that Oppenheimer was, to put it mildly, not enthusiastic. Then Truman asked Oppenheimer when he thought the Russians would get the bomb. Oppenheimer said he did not know. Truman responded by saying that he did, "never." Then Oppenheimer volunteered a remark which ended any hope of any relationship with Truman. He said, referring to the bomb, and wringing his hands, "I have blood on my hands." Truman went livid and threw Oppenheimer out of his office saying afterwards that he never wanted to see "that son-of-a-bitch" in his office again. Truman was the one who had ordered the bombing of Hiroshima and Nagasaki. I believe, and will explain, that this exchange had profound consequences for the future of nuclear weapons.

The people associated with the creation of the bomb speculated as to when the Russians would have one. General Groves, who headed the project, said twenty years. Some people thought two or three. The consensus among the physicists was for about five. The intelligence estimate was for 1953. There was no question among the physicists that the Russians would build one. One of the reasons was that they had an appreciation for the level of Russian physics. For example, when it became clear in the 1930's that Victor Weisskopf, (who became second in command of the Theory Division at Los Alamos) that as a Jew he was going to have to leave Europe, he considered an offer from the University of Kiev (he had spent a fair amount of time in Russia) but decided against it and went instead to the University of Rochester. He knew a great many of the Russian physicists. The brilliant Russian theoretical physicist Lev Landau (later a Nobel Prize winner) came and went from Niels Bohr's institute in Copenhagen. An interesting case is that of Pytor Kapitsa—another Nobel Prize winner. He got his education in Russia but then came to Cambridge where he worked for a decade, even directing a laboratory there. In 1934, he returned to Russia to visit his parents and was not allowed to leave. The people at Los Alamos probably speculated correctly that he would be involved in any attempt to build a bomb. Also, the Russians had published important work in fission.

Heavy nuclei, such as those of uranium and plutonium, contain so many neutrons and protons that following them individually is impractical. Thus one studies their collective behavior. For many purposes, this behavior is like that of a liquid drop. There is a surface tension that holds these nuclei into a roughly spherical shape. If a neutron impinges on such a nucleus, it can cause the "drop" to oscillate and break apart—to fission. The products are nuclei in the middle of the periodic table and neutrons which can cause further fissions. But these heavy nuclei are unstable and can also fission spontaneously. This effect was first observed by two Russian physicists, Georgy Flerov and Konstantin Petrzak in 1940, in experiments they did in the Dinamo Station of the Moscow subway. They chose the underground to shield away cosmic rays. This result was well-known to the nuclear physicists in Los Alamos but what was not well-known, at least at first, was what effect this would have on the design of nuclear weapons. To understand this, we must know a bit of how plutonium is produced in a nuclear reactor.

Reactor fuel is largely uranium 238. But if such a nucleus absorbs a neutron it can become uranium 239 which decays into neptunium 239 which in turn decays into plutonium 239. Thus plutonium is produced in nuclear reactors. Some atomic bombs such as the Hiroshima bomb, have as their essential fuel uranium 235. In the Hiroshima bomb, a mass of uranium less than a critical mass is shot into another sub-critical mass. It takes the order of milliseconds to assemble a critical mass. However, Uranium 235 can fission spontaneously and if too many neutrons are produced before a critical mass is assembled, there will be pre-detonation—a "fizzle" to use the term of art. The spontaneous decay rate for U235 is sufficiently small so that this is not a problem. The same thing is true for plutonium 239. But when the first samples of plutonium were delivered to Los Alamos in the spring of 1944, it was discovered that they had a high rate of spontaneous fission. This is because they were "contaminated" with plutonium 240, which was produced when plutonium 239 absorbed a neutron. The gun assembly method was ruled out. Instead a sphere of plutonium had to be compressed—"imploded" by the use of high explosives thus increasing its density and lowering the critical mass. It turns out that this can be done in microseconds solving the spontaneous fission problem. This is something that the Russians would have had to discover for themselves, except that the way they discovered it was by espionage.

As far as we know, there were three Soviet spies at Los Alamos, all unknown to each other as spies. David Greenglass was a machinist who worked on shaping the high explosives used to implode the plutonium. He delivered somewhat crude drawings of these explosive "lenses" to the Russians. Ted Hall was nineteen when as a fresh Harvard graduate, he came to Los Alamos in 1944. He supplied some material to the Russians but his main importance was to confirm some of the material that the real Russian spy Klaus Fuchs had delivered. The Russians were suspicious that what they were being delivered was a plant designed to throw them off. These confirmations were important. Even more important was the judgment of the very few Russian physicists who were allowed to see this material. They knew that what they were seeing was priceless.

Klaus Fuchs was born in Rüsselsheim, Germany in 1911. His father was a Lutheran pastor, who taught at the University of Leipzig which Fuchs attended. Fuchs became involved in politics and joined the student branch of the Social Democratic Party of Germany. Fuchs came to side with the communist wing of the party and was expelled. He joined the Communist Party in 1932 and then was forced to leave Germany. He took exile in England and got his PhD in physics from the University of Bristol. He then did postdoctoral work in Edinburgh with the great German physicist Max Born, also a German refugee. In 1940, Fuchs was classified as an "enemy alien" and sent to Canada. Rudolf Peierls, who with his fellow refugee Otto Frisch wrote the memoranda that started the British atomic bomb project, arranged for Fuchs to return to England and join him in the project. Peierls was then at the University of Birmingham and he and his wife Genia often took in physicists as borders. Fuchs came to live with them. Genia complained that he never said anything and compared him to a music player which only produced a sound when it was fed a coin. The British had decided to go the way of separating uranium isotopes and Fuchs and Peierls wrote a seminal paper on the subject. They concluded that the then available centrifuge technology was too primitive and focused on the diffusion of a uranium gas. The gas if forced through tiny pores, the lighter isotope diffuses first allowing the separation. By 1942, Fuchs had made arrangements to transmit information to the Soviet Union. His plans were interrupted when Peierls was transferred to Columbia University in New York to continue with this work. The odd thing was that Fuchs did not want to go from Columbia University to Los Alamos. He wanted to go back to England. A remarkable letter which I will now quote exists from James Chadwick, the discoverer of the neutron and the head of the British delegation. It was written to Peierls. There are some names of British physicists we don't need to identify. "Bethe" is Hans Bethe who was the leader of the Theory Division at Los Alamos, which is referred to here by its code name "Y".

July 14, 1944

*"Dear Peierls,*

*I have now had talks with both Kearton and Fuchs about the future of the New York section and in particular about their own positions. As a result, Kearton will approach Keith and Benedict with the object of getting a letter by one or both of them to Groves to say that the services of Fuchs and Skyrme are no longer required. It is possible that this matter was raised by Groves on a visit to New York earlier in the week, but I have had no news from him so far.*

*The position of Skyrme is quite clear. Bethe or Oppenheimer should write to Groves asking for his services in Y. Groves has provisionally agreed and there should be little delay over his transfer.*

*Fuchs' future is not so clear. I gave you the gist of a cable from Akers in my letter of July 11. I did not agree with the suggestion made in this cable that Fuchs was not required in England, but I wished to discuss the question with Kearton before I made up my mind. Kearton was very strongly of the opinion that Fuchs was quite necessary in England if work on any kind of diffusion plant is to continue...*

*I have now had a talk with Fuchs himself. He feels that he has a special contribution to make in England, whereas in Y he would be one of a number and can make no really significant difference to the work*

I agree completely with these views of Kearton and Fuchs, and I feel sure you also agree at least in principle.

I come now to the point of this letter It would put me in a very awkward position if a request for Fuchs' services in Y were to be sent to Groves. If Groves were to agree I also should have to consent, for the consequences of refusing, on the grounds that he was needed in England for work which can have no significance for the war, might be quite serious. It would certainly cause great resentment in some quarters and our relations with the U.S. on this project would be impaired. I should attempt to justify his return as being useful for the New York project, for after his experience here he could interpret their requests and help to direct U.K. work into directions of immediate interest to them. This argument would of course not be valid if a low-separation diffusion plant were to be started in England.

I therefore do not want Bethe to ask for Fuchs. Further than that, I want Bethe to say that Fuchs would not be specially useful in Y, if Groves asks if they want him, as he may. This means some tactful work on your part and I hope you will be able to do what is necessary by suggestion rather than direct action.

I have prepared the ground here and I think the matter can be arranged. I have stated that Fuchs could be useful in Y but that his special qualifications are not on the nuclear side but on the diffusion plant.

Until I know something of what is happening in London I want to keep the New York position as fluid as possible.

Yours sincerely,
J. Chadwick"

Somehow, the matter was resolved and Fuchs joined the Theoretical Division at Los Alamos in August of 1944. On a visit to his sister in Boston in February of 1945, Fuchs made his first transmission regarding Los Alamos data to his courier Harry Gold. What has always puzzled me was in what form the information was taken out of Los Alamos. This transmission was relatively brief but later there would be multiple pages. He certainly turned over these transmissions to Gold in written form since Gold would have no understanding of them. I have been told that the security at the gate in Los Alamos was somewhat erratic. In fact, people left the laboratory with secret documents that they had forgotten to leave behind. But I doubt that Fuchs would have taken this risk. If he had been caught, he would have been finished. He had a photographic memory and I think that it is more than likely that he put what he remembered into written form after he left the laboratory. In this case he informed the Russians about the discovery of spontaneous fission in plutonium and the switch to implosion. Hall had previously told the Russians about implosion in less detail.

Until the use of the bombs on Hiroshima and Nagasaki, the Russian program had been somewhat casual. But now Stalin ordered a crash program to build the bomb. This was preceded by a detailed report by Fuchs on the plutonium implosion device which he transmitted in June of 1945. The Russians have now released this and I give a sample below.

**Active Material**

The element plutonium of delta-phase with specific gravity 15.8 is the active material of the atomic bomb. It is made in the shape of a spherical shell consisting of two halves, which just like the outer spherule of the initiator, are compressed in a nickel-carbonyl atmosphere. The outer diameter of the ball is 80-90 mm. The weight of the active material including the initiator is 7.3–10.0 kg. Between the hemispheres is a gasket of corrugated gold of thickness 0.1 mm, which protects against penetration of the initiator by high-speed jets moving along the junction plane of the hemispheres of active material. These jets can prematurely activate the initiator.

In one of the hemispheres, there is an opening of diameter 25 mm, which is used to insert the initiator into the centre of the active material, where it is mounted on a special bracket. After inserting the initiator, the opening is closed with a plug, made also of plutonium.

One of the most significant revelations in this document has to do with the $\delta$ phase of plutonium. After Glen Seaborg and collaborators first observed plutonium made in a cyclotron in 1941, micrograms became available for experimentation. William Zachariasen, a crystallographer working at the University of Chicago found that the element existed in several "allotropes." These are different physical forms such as diamond and graphite for carbon. The $\alpha$ phase is stable at room temperature but is a chalk and useless for making a bomb. The $\delta$ phase is metallic but unstable at room temperature. It had to be alloyed with gallium for it to work. Being told that the plutonium used to make a bomb was in the $\delta$ phase saved the Russians a great deal of time. As did the rest of what Fuchs had told them. Once they decided to make a bomb they created an organization to do it. It was headed by Lavrenti Beria, Stalin's much feared chief of the secret police. The scientific director was the physicist Iulii Borisovich Khariton. He had access to all of the espionage data which he decided to confirm independently. He was ordered by Stalin to produce a clone of the plutonium bomb that flattened Nagasaki. Kapitsa contacted Stalin to suggest alternate ways of doing the job. He was summarily thrown off the project and placed under house arrest where he remained until Stalin's death. The physics was done in a new facility which had been built in the Urals town of Arzmas which the scientist came to call Los Arzmas. Since they had decided to go the plutonium route they had to build reactors. The first one went critical in December of 1946. They also created a vast mining industry to mine uranium. The successful test of what we called "Joe-1" occurred at 7a.m. on August 29, 1949. It had about the same explosive yield as the Nagasaki bomb on which it had been modeled.

President Truman's first reaction was that it was a fake—that those " Asiatics" could never build an atomic bomb. But once the radioactive fall-out drifted to where it could be observed it was clear even to

Truman that those "Asiatics" had detonated a bomb. On September 23, he made an announcement.

"We have evidence that within recent weeks an atomic explosion occurred in the U.S.S.R

Ever since atomic energy was first released to man, the eventual development of this new force by other nations was to be expected. This probability has always been taken into account by us.

The recent development emphasizes once again, if such emphasis was needed, the necessity for that truly effective, and enforceable international control of atomic energy which this government and a large majority of the United Nations support."

Whatever good intentions Truman might have had in this declaration, they were rapidly overwhelmed by the idea that they were in an existential struggle with the Russians and that they must stay ahead of them in the nuclear arms race. As I will now discuss, this was what led Truman on January 30, 1950 to publicly announce that the United States was going to initiate a crash program to build a hydrogen bomb, something which no one at the time knew how to do and about which Truman had no real understanding. It was a terrible decision. If Truman had paid attention to the report of a scientific advisory committee headed by Oppenheimer it might have been avoided.

A hydrogen bomb is so named because its energy source is the fusion of isotopes of hydrogen, a process that can release energy. The most energetically productive reaction is the fusion of a heavy hydrogen nucleus which consists of one neutron and one proton with a nucleus of super heavy hydrogen which consists of one proton and two neutrons, to produce a nucleus of helium and a neutron that carries away most of the energy. To initiate such a reaction, one needs an atomic bomb to produce the requisite temperatures. The first suggestion of doing this seems to have come from the physicist Enrico Fermi. He had used his trip to Sweden to collect his Nobel Prize, to leave Fascist Italy and come to the United States. Early in 1942, Edward Teller visited him at Columbia

University where he had become a professor to discuss atomic weapons. Fermi casually suggested that they might be used to create conditions of temperature that pertain in the interior of stars, where fusion reactions provide the energy. Teller at first thought that the idea was crazy but he soon became obsessed by it. He thought that the fission device was as good as done and that the focus should be on the fusion device. This obsession never left him. During the war, Oppenheimer had to create a special division at Los Alamos, in which Teller and a small group could work on the "super." They found themselves stuck. You could ignite some portion of the fusible material but then it would radiate and cool off before the fusion reactions could propagate. It was like trying to light a log with a match. In early September of 1945, Fermi gave some summary lectures on the super. They read like a physics text. He argued that he did not see how to make it work. One of the members of the audience was Fuchs.

On September 19, the British delegation gave a formal lab party. Fuchs was in Albuquerque where he had gone to meet his Russian courier Harry Gold. He turned over notes on Fermi's lectures. Again I do not know in what form. One can find the lectures on the internet but they seem to be a later version. This was the most significant thing that Fuchs gave the Russians on the hydrogen bomb. It revealed both the attempt to fuse heavy and super heavy hydrogen but also the difficulties. Above all it revealed that we had a program and the Russians began looking into the matter.

In 1946, Truman signed an act that converted atomic energy from a military to a civilian domain. The Atomic Energy Commission was created—now replaced by the Department of Energy. In October of 1949, a General Advisory Committee was asked to make recommendations about constructing a hydrogen bomb. Oppenheimer was the chairman and the committee consisted of people like the physicist I. I. Rabi and Enrico Fermi, as well as the president of Harvard James Conant. Some years ago, Rabi described for me how it went.

He said, "*In fact, the debate was recorded but the tapes were deliberately destroyed soon afterward. It is a great pity that we cannot hear the voices of Fermi and Oppenheimer and the rest in that very fateful discussion…*"

"*… after the Russian explosion, certain people thought that we need some counterthrust. And [Ernest] Lawrence, [Luis] Alvarez and Teller felt that the only thing was to go full tilt for the Super. In fact, during Los Alamos some people felt that it should have higher priority than the ordinary fission bomb. 'We shouldn't dillydally with the fission bomb but go for the Super'—so said some very eminent physicists. Not very sensible, but very eminent. I wonder what they think about that now. Anyway after the Russian explosion they wanted to go ahead with a specific Super model. This model is what came up before the committee. There were two objections to it. In the first place it was a very chancy thing because, basically, we did not know how to make it. And then, just about this time it was shown that it wouldn't work, it wouldn't propagate-at least not that particular model. But the general kind of thing they were talking about would have been absolutely devastating because it was so big.…*"

"*In any event, there was strong agreement within the committee that we should not go ahead. We all agreed that if it could be made to work it would be a terrible thing. It would be awful for humanity altogether. It might give this country a temporary advantage, but then the others would catch up-and it would just louse up life. Fermi and I said that we should use this as an excuse to call a world conference for the nations to agree, for the time being, not to do further research on this.*"

Truman decided to bypass the scientists and bring in his own advisors, three non-scientists, and as Rabi noted, "*…it was just about this time that Klaus Fuchs was arrested in London. All of this got to President Truman and built up such a head of steam, that he was practically forced to declare that he was going to give the Super top priority.…*"

*I never forgave Truman for buckling under pressure. He simply did not understand what it was about. As a matter of fact, after he stopped being president, he still didn't believe that the Russians had a bomb in 1949. He said so. So for him to have alerted the world that we were going to make a hydrogen bomb at a time when we didn't even know how to make one, was one of the worst things he could have done. It shows the dangers in this kind of thing. He didn't have his own scientific people to consult and give him an impartial picture."*

One is led to wonder that what if Fuchs had not given the Russians what he did? Perhaps they would not have gotten the bomb until Truman left office and might Eisenhower have taken a different route? In 1951, the Polish born mathematician Stanislaw Ulam and Edward Teller proposed a novel design of a fusion bomb. In essence, it was a two-stage device, in which as in the classical super, the first stage was a fission bomb. But the radiation emitted by this explosion was used to compress a container of the fusible material thus causing it to ignite. Using this design our first successful test of a hydrogen bomb was in 1952. The Russians followed in 1955, the British in 1957, the French in 1968 and the Chinese in 1967. At first, the idea was to make bigger and bigger bombs. Here is a useful scale. In 1995, Timothy McVeigh brought down half of the Alfred P. Murrah Federal Building in Oklahoma City with a Ryder Truck packed with 5,000 pounds (2.5 tons) of high explosive. Here is the number of Ryder trucks equivalent to these bombs. The bomb that brought down Nagasaki is about 8,000 Ryder trucks. Hydrogen bombs correspond to millions of trucks. The biggest test was the Russian Tsar Bomba in 1962. It was about 3,000 times more powerful than the Nagasaki bomb. But it was realized that there was another use for staged nuclear explosions. One could drastically economize on the plutonium used in fission bombs. The bombs could also be lighter, which is helpful if they are to be put in rockets and they would be much more powerful than say the Nagasaki bomb. They could be built in the thousands, which is what has happened in the United States and Russia. What strikes me about all of this is that it is pointless. The targets are still too small. There are asymmetric wars all over the world. Can anyone imagine a role for any of these bombs? Human ingenuity has managed to create the most devastating and the most useless weapon ever invented.

## 14
## Round and Round

At the beginning of the Second World War, the great British theoretical physicist P. A. M. Dirac was asked to work on the nascent British program to build a nuclear weapon. It was understood from the beginning that the key to the development was the production of the explosive fissile fuel. The first choice was the lighter isotope of uranium (uranium 235) which was found in concentrations of less than a percent in natural uranium, which is mostly uranium 238. Dirac had been interested in the separation of isotopes as sort of an avocation since the 1930's. Hence he began work on this subject for the nuclear program. Being Dirac, he approached this matter with great generality. He did not concern himself with specific methods of separation, such as centrifuges, but with general principles that would govern all methods. To this end, he invented a unit separation, the SWU—pronounced "swoo." The detailed definition of this unit is very non-intuitive. It would have taken a Dirac to have found it. But its uses are very clear. You can ask how many SWU in a year does some given centrifuge produce and you can ask how many SWU are required to start with natural uranium hexafluoride and produce a kilogram of enriched uranium hexafluoride. This enriched uranium consists of a mixture which has a concentration of 95% uranium 235 (weapons grade). The answer is about 232 SWU. We can now turn to the production of SWU by centrifuges and in particular, to the program in Iran.

The Iranian program began in 1993, with the sale to Iran of both a sample model and the plans for a centrifuge which had been used in the Pakistani nuclear program. The salesman was the Pakistani proliferator A. Q. Khan. To appreciate what the Iranians did, one must emphasize that a single centrifuge is useless. It might produce a few SWU per year. What is necessary is to manufacture thousands of centrifuges and array them in cascades. This is a very non-trivial matter. In the event, Iran was able to deploy over 19,000 centrifuges. Most of these are of the earlier type, with a limited SWU production but still there are enough of them to manufacture weapons-grade uranium for nuclear weapons in a fairly short order. In addition, Iran began manufacturing more advanced centrifuges. The most advanced, the so called IR-8 centrifuge, is said to be able to produce about 24 SWU per year, or about ten kilograms of highly enriched uranium. With 25-30 kilograms of highly enriched uranium, one can make an implosion-type nuclear bomb with a yield on the order of ten kilotons. And with even a little less, one can make a bomb with a yield exceeding one kiloton. The Hiroshima bomb had a fifteen kiloton yield. Thus if these centrifuges were deployed, the fuel for nuclear weapons could be manufactured very rapidly. On the positive side it takes perhaps a decade to produce and deploy such centrifuges but there are IR-6 centrifuges some of which have been deployed,

The people who created the nuclear treaty with Iran were very well aware of this. The treaty stipulated that the approximately 19,000 installed centrifuges should be reduced to about 6,000. Five thousand of these can be used to enrich uranium for reactor fuel. The deployed IR-6 centrifuges are not supposed to be used for uranium enrichment but the language appears ambiguous and the Iranians are talking advantage of that to sail close to the wind. One thing should be very clear. If the treaty is abolished and the IAEA inspections stopped there will be little to halt the Iranians in a rapid quest for weapons grade uranium. Trump appears to have no understanding of this at all, He seems to think of this as a commercial "deal" in which we "gave" the Iranians money making them "rich" and got back "nothing". What we got back was at least a postponement of what could be a real disaster both for the Iranians and everybody else.

# 15
Li6

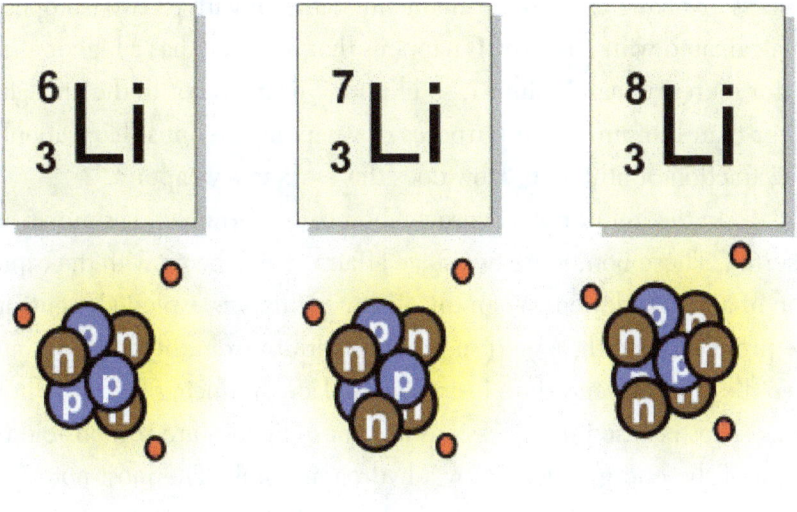

Lithium 6

Lithium 6 and lithium 7 are the two stable isotopes of lithium left over from the Big Bang. In nature, lithium 7 has nearly a 93% abundance. Unfortunately, lithium 6 has one application and that is in nuclear weapons. This comes about because of nuclear fusion. Energy is produced when two light elements such as the isotopes of hydrogen fuse. The most energetically favorable reaction is the fusion of heavy hydrogen (the deuteron) with super heavy hydrogen (the triton) to produce helium and a very energetic neutron. In symbols,

$$D + T \rightarrow He + n.$$

While heavy hydrogen can be separated from ordinary hydrogen by standard processes, tritium (the T) is produced in a nuclear reaction involving lithium 6. Lithium 6 absorbs a neutron to become a triton plus a helium nucleus. But how to produce lithium 6? This is done by a process involving chemical enrichment. Lithium with its two isotopes is amalgamated with mercury. It happens that lithium 6 has a higher affinity for mercury than lithium 7, so lithium 7 diffuses out of the amalgam faster than lithium 6. For purposes of weapons one must have about a 95% fraction of lithium 6. How does this work in a weapon?

In the interior of a hydrogen bomb, deuterium and lithium 6 are inserted. These bombs are two stage affairs, which begin with the explosion of a fission device. When this fission primary is exploded, neutrons are produced and these in turn produce tritium from the lithium 6, and then the tritium-deuterium fusion takes place producing energetic neutrons. These cause further fissions and these fissions are responsible for much of the energy released by a hydrogen bomb. The most notorious case was the 1954 Castle Bravo hydrogen bomb test in the Pacific in which the yield was underestimated because one did not take fully into account the role of lithium. Here what was not accounted for was lithium 7, which will always be present since the lithium 6 used is never pure. When lithium 7 is hit by a neutron, it can be split into a helium atom and a triton plus a neutron. This produces energy and must be taken into account. The Japanese crew of the unfortunately named Lucky Dragon were irradiated. When one observes a country arranging to manufacture lithium 6, it can only mean one thing—nuclear weapons. This is precisely what seems to be happening in North Korea. In a recent report by the Institute for Science and International Security, David Albright and collaborators give the details.

They note that in 2012, North Korea was able to procure wherewithal to construct a mercury lithium 6 enrichment plant. They cannot pin down the precise location of this plant, which once again illustrates the elusive and dispersed nature of their weapons program. Research on ways to produce lithium 6 have been published by the North Koreans in the open literature. It seems as if the last nuclear device tested by the North Koreans was a boosted nuclear weapon involving both fission and fusion. These boosted weapons use tritium gas and these tritons are manufactured using lithium 6. This could be a first step towards making a full scale hydrogen bomb, in which lithium 6 will play a direct role. There is no other reason for manufacturing it.

# 16

## Is E = mc²?

The purpose of this note is to persuade you that the formula $E = mc^2$ has never been proven nor has it been tested. What has been proven and tested is the formula $\Delta E = \Delta mc^2$, where $\Delta$ is the "change in". I will begin, by giving you a version of Einstein's 1905 argument which he might have given, if he had taken his own presentation of light quanta in his 1905 paper on the photo-electric effect seriously. He never considers the momenta of the quanta in the photoelectric paper, since all he is concerned with is the conservation of energy. The first time the momenta are taken seriously is in the kinematics of Compton scattering. Then I will give you a formal argument of the kind you may have been given in an introductory course on special relativity. A side note, I will spare you Einstein's notation. Until 1907, he called the speed of light capital V, which makes these early papers a bit of a challenge to read.

Let us suppose, that there is a body of mass m, at rest, that emits two light quanta of energy hv and momenta hv/c and –hv/c. We now wish to observe this from a frame of reference moving with speed v and at an angle $\Theta$ from the x axis. We can apply the relativistic Doppler shift

$$v' = \frac{v}{\sqrt{1-\frac{v^2}{c^2}}}\left(1 + \frac{v}{c}\cos(\vartheta)\right) = \gamma v(1 + \beta \cos(\vartheta)).$$

Thus

$$v'_1 = \gamma v(1 + \beta \cos(\vartheta)).$$

And
$$v'_2 = \gamma v(1 - \beta \cos(\vartheta)).$$

I will now follow Einstein since he seems to have gotten the correct result. We apply the conservation of energy in both frames. I will call L = h$v$ and the subscript 0 stands for before the photon emission and the subscript 1 for after the photon emission. I will again be following Einstein, to call the rest frame quantities E and the moving quantities H.

Thus after the suitable subtractions and invoking the conservation of energy,
$$E_0 = E_1 + L$$
and
$$H_0 = H_1 + L\gamma.$$
So
$$H_0 - E_0 - (H_1 - E_1) = L(\gamma - 1).$$

Now comes Einstein's bit of magic. He asks how can H and E differ. He argues that it can only be by the kinetic energy K, which is defined up to an arbitrary constant. He then concludes that the above expression represents the change in the kinetic energy, which can come about only by a change of mass. Einstein evaluates this equation for small v/c. He concludes that if the body emits photons of energy L, this corresponds to a loss of mass $L/c^2$.

There was some contemporary criticism of this result, because of the small v/c approximation. I will give you a standard formal derivation where this objection does not arise. But pay attention to what has been alleged here. Nowhere is it stated that $E = mc^2$. Rather, a change in E is proportional to a change in m. I do not know of any derivation which proves more than this. Furthermore, no experiment proves more than this. I invite you to go through the list from fission on down. The derivation I will now give you is an exercise in differentiation. We begin with the equation for energy in terms of work.

$$dE = Fdx = (dp/dt)dx = dp.v$$

and the relativistic definition of p,

$$p = m_o \gamma v = mv.$$

Thus

$$dp = dmv + dvm$$

or

$$dE = mvdv + v^2 dm = c^2 dm.$$

This last is just differentiation. Now what to do? If we integrate we get a constant. Here, Pauli in his wonderful monograph has a splendid suggestion.[1]

"One could think of determining the constant in such a way that $E_{kin}$ is zero for a particle at rest. It is however more practicable to put the constant itself equal to zero."

This term is small because I neglect it.

Shortly after Einstein settled in Princeton, he gave the Eleventh Josiah Willard Gibbs lecture in Pittsburgh. It was entitled "Elementary Derivation of the Equivalence of Mass and Energy." It is available on the site I reference below. At the end, he rejects all derivations such as the one I just gave that make use of the notion of "force." He writes, "Thus in the book just mentioned [a book on relativity], essential use was made of the concept of *force* which in the relativity theory has no such direct significance as it has in classical mechanics. This is connected with the fact, that in the latter, the force is to be considered as a given function of the coordinates of all the particles, which is obviously not possible in the relativity theory. Therefore, I have avoided introducing the force concept."

What strikes me about this paper is its relative mathematical sophistication as compared, say, to his 1905 paper. The first equation is

$$ds^2 = dt^2 - dx^2 - dy^2 - dz^2.$$

When Hermann Minkowski produced his four dimensional formulation of relativity, Einstein initially rejected it as an irrelevance. He once remarked that since the mathematicians had gotten a hold of his theory, he no longer understood it himself. Einstein's view completely changed once he began formulating his general theory of relativity. What is curious about this paper, is that he explicitly rejects the use of the four dimensional formulation as an unnecessary complication once he has introduced it.

He considers the collision between two particles of mass m in the center of mass and the laboratory frames. He then derived the relationship for before and after the collision, using somewhat different notation than Einstein and putting in the speed of light explicitly

$$E'_0 - E_0 = m'c^2 - mc^2.$$

Here $E_0$ is what Einstein refers to as the "rest energy." The total energy is $E_0$+kinetic energy. He notes that the rest energy is only determined up to an additive constant. He then concludes that

"If for collisions of material points conservation laws are to hold for an arbitrary (Lorentz) coordinate system, the well-kno.wn expressions for impulse and energy follow, as well as the validity of the principle of equivalence of mass and rest energy."

Whether this derivation is more transparent than my modified version of Einstein's 1905 derivation is in the eye of the beholder. In any event it is clear that what is generally valid is the equation

$$\Delta E = \Delta mc^2.$$

## Endnote

1. Theory of Relativity, Dover p. 116

# V Life

## A Trick of Memory 17

I started what turned out to be a two year visit to the Institute for Advanced Study in Princeton, in the fall of 1957. I had gotten my PhD at Harvard in 1955 and then had worked for two years as the "House Theorist" at the Harvard Cyclotron. At the end of this time a revolution had occurred in the field of physics that interested me — elementary particles. Up to this time it had been felt that descriptions using left handed or right handed coordinate systems were equally valid. But experiments suggested by the Chinese American physicists, T. D. Lee and C. N. Yang showed that for certain processes, the choice mattered. Both Lee and Yang were at the Institute and that fall, they won the Nobel Prize. Eventually I had the chance to work with them. This was where the excitement was. But somehow in the middle of this, that fall a

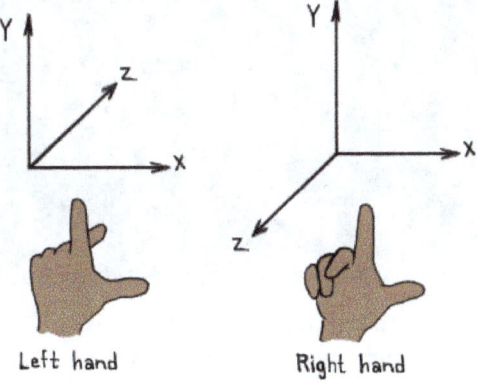

Left and right handed coordinate systems.

rumor came from Europe that Werner Heisenberg and Wolfgang Pauli had made a profound theoretical advance. I can't recall how we learned this. I do not think there was a paper. Maybe Robert Oppenheimer, our director, had gotten word. He always had his nose in the wind. The reaction was mixed. Heisenberg made various declarations from time to time at this period of great advances and I don't think that much attention was paid. But Pauli was something else. He was a scathing critic of wrong or trivial work. Of one paper, he said that it was not "even" wrong. Of physicists he did not admire, he said, "So young and already so unknown." This latter reflected his Viennese roots. In Die Fledermaus there is a masked ball and when a maid in disguise is introduced to a prince she says, "So young and already a prince." Once it was known that Pauli was involved, attention had to be paid.

The Pupin Lecture Hall. If you look carefully you can see Dyson and me on the right side about half way up.

The sequence of events I am going to describe began on Thursday, January 30, 1958 when Pauli arrived in New York by boat. Word had come to the Institute that he was going to give special colloquium, open only to invitees at Columbia University. Since he was leaving New York for Chicago and Berkeley, on the second of February, the colloquium was for either the $31^{st}$ or the $1^{st}$ of February and I do not remember which. But a delegation from the Institute went to New York for it. The senior members included Oppenheimer, Bohr, Lee, Abraham Pais and Freeman Dyson. I sat next to Dyson somewhere in the middle of the auditorium which is pictured above. The rest of these dignitaries were in the first row. I do not remember anything of Pauli's talk, but I very distinctly remember the aftermath. Pauli went over to Bohr wanting to know his views. He said to Bohr that the theory might look "crazy." Bohr said that the problem was that it was not crazy enough. He meant that a really profound theory like relativity or quantum mechanics always looks crazy in the beginning. Pauli responded by saying it was crazy enough. There then ensued something that no dramatist could have invented. Bohr stood up and two of them began following each other around the table in the front of the lecture room. When Pauli got in front, he said the theory was crazy enough and when Bohr got in front, he said it wasn't. I thought that if someone from the outside world witnessed this performance from two of the greatest physicists of the $20^{th}$ century, they would think that we are all crazy. At one point, Dyson said to me of Pauli's performance that it was like watching the death of a noble animal. In fact, not long afterwards Pauli thought the better of the whole thing and he sent Oppenheimer a little drawing which he said had been done by Heisenberg. It was an empty canvas, with caption underneath that read, "I can paint like Titian, only a few details are missing."

Wolfgang Pauli and Niels Bohr and the Tippe Top[1]

I bring all of this up now, because I want to discuss the tricks of memory. I carry on an email correspondence with several of my colleagues about this and that. For reasons that I cannot explain, I decided in one of my emails to recount the story I have just given. I thought it would be entertaining. I was astonished to receive the response I did from Steve Weinberg. He reminded me that at the time, he had been a young instructor at Columbia and that he too had a vivid memory of this occasion, but there was one thing that I had gotten wrong. The talk was not given by Pauli but by Heisenberg! Steve is one of the smartest people I know and his email was very disturbing to me. My memories of Pauli are so vivid, how could I possibly have made such a mistake and how could I now, after fifty-eight years, reconstruct the event? Most of the people who were there are long gone — Pauli, Oppenheimer, Bohr, Pais and so on. Who could I ask

---

[1] Photo credits: Emilio Segre Archives

and how reliable would their memories be? I began emailing people. Richard Garwin was there but he was not sure whether it was Pauli or Heisenberg. Freeman Dyson and T. D. Lee, while not recalling much about the event, remembered it as being Pauli. Steve remained unconvinced. I then thought that I would consult Heisenberg's biographer, David Cassidy, who is a meticulous gatherer of facts. He turned up a letter written by Pauli to Heisenberg, dated February 1. From the letter it seems clear that the talk was given on Friday the 31$^{st}$. In what he wrote, he mentions an "informal lecture" he had given and then says, "Bohr was present (I saw him only briefly), and he had reservations that our basic assumptions (like indefinite metric) are not really radical enough in order to encompass enough new elements for the solution of all our problems." This persuaded Steve and he has no idea how this trick of memory occurred. Today, this trick of memory is known in online circles as the Mandela Effect. Speaking of tricks of memory, Freeman Dyson wrote

> This is indeed a good lesson for us all. Here is another example. I recently wrote an account of a dinner-party at the Oppenheimers' home fifty years ago. I remembered the details of the evening clearly and wrote a vivid description of it. The Kennans and the Niebuhrs were there, and Robert Oppenheimer read aloud poems of George Herbert that he loved.
>
> After I finished writing the piece, I decided to triangulate. I looked up an old letter from Ursula Niebuhr and found in it a description of the party written at the time. All the details in my memory were correct, except for one. I was not there. Somehow my memory had transformed Ursula's story into a personal experience.
>
> Yours sincerely,
> Freeman Dyson.

# Checkers  18

The fireplace at Fine Hall with the aphorism above.[1]

Checkers

First published in the London Review of Books April 27, 2017

Sometime in the early 1970's I got the idea that I wanted to write a profile of Albert Einstein for the New Yorker. I had been writing for them for about a decade so at least I had an inside track. Furthermore, there were almost no biographies. My teacher Philipp Frank, who had known Einstein for decades, had written one, *Einstein: His Life and Times*, which I had read soon after it came out in 1948. I had no idea

---

[1] Photo credits: Courtesy of the Archives, California Institute of Technology

what I could add to Professor Frank's biography or even how I could begin one of my own but I did have an ace in the hole — Helen Dukas. Miss Dukas had come with Einstein as his secretary when he emigrated from Germany. Miss Dukas became a member of the family and lived in the Einstein house in Princeton which after his death she shared with Einstein's step — daughter Margot. I had no idea what I would find but I asked Miss Dukas, whom I had known since my days at the Institute for Advanced Study if I could visit the house. She agreed. She showed me Einstein's upstairs sort of apartment quarters. It still had some of the books such as Frazer's The Golden Bough which he read. I noticed the wall decorations. There was an etching of James Clerk Maxwell and one of Newton which had come out of its frame. This seemed to me to be symbolically correct. Einstein's theory of gravitation had replaced Newton's but we still had Maxwell's electrodynamics. While this was interesting I still did not see how to proceed. Helen offered to make lunch and while she was preparing the sandwiches she gave me a book to look at which had to do with some of Einstein's writing on peace. Most of it seemed to me to be sort of uninteresting but then I came across his correspondence with the Queen of Belgium. After reading the first letter dated in 1930 which was addressed to his wife Elsa and read,

> I went across to the station to telephone the Kings. (King Albert was still alive although he was killed in a mountaineering accident in 1934.) It was quite tedious because the line was always busy…At 3 o'clock I drove out to the Kings, where I was received with touching warmth. These people are of a purity and kindness seldom found. First, we talked for about an hour. Then an English woman musician arrived, and we played quartets and trios (a musical lad-in-waiting was also present). This went on merrily for several hours. Then they all went away and I stayed behind alone for dinner with the Kings — vegetarian style, no servants. Spinach with hard-boiled eggs and potatoes, period. (It had not been anticipated that I would stay.) I liked it very much there, and I was certain that the feeling was mutual.

I immediately understood how to begin my profile.

Indeed, seeing these letters aroused in me the sort of feeling a prospector might have if he came upon a gold nugget. I had no idea what the next step was going to be but this was certainly the first. Much later I learned that Leo Szilard had asked Einstein in the beginning of the war to ask the queen not to let the Germans export uranium from the Belgian Congo. It took many months to do the writing and a long time to do the editing which was done by Pat Crow — a monumental job. Then came the checkers. The New Yorker checkers were and I think still are legendary. They are there to help the writers although a few like Jay McInernany who was fired think that they should be doing the writing. I hit on one of those. Most of it went smoothly until we came to the famous Einstein aphorism *"Raffiniert ist der Herrgott, aber boshaft ist er nicht."* Which I would translate as God is sophisticated but not malicious, but some like "subtle" rather than "sophisticated."

I had a long history with the aphorism. There was a physicist named Max Herzberger who had come as a refugee in the 1930's to Rochester, New York where I grew up. He was employed by Eastman Kodak as lens designer. Einstein had been one of his PhD examiners in Berlin and from time to time he went to visit him in Princeton. Long before I studied physics Max told me this aphorism which I loved and never forgot. I had no idea of a source other than Max and the checker would not let me use it until I could produce a source. I did not even know when Einstein had said it or why. In desperation I called the Princeton mathematics department. I knew that when Einstein came to Princeton the Institute for Advanced Study did not have facilities and Einstein had his office in Fine Hall where the math department was so that if he had ever visited Princeton earlier he would have come to Fine Hall. I spoke to a department secretary and she said "Of course. It is over the fireplace outside my office in the German."

But why did it get there?

In 1921 Einstein made his first visit to Princeton. He was informed that a physicist named Dayton Miller at the Mount Wilson Laboratory in Pasadena had just done an experiment which if correct would have invalidated the theory of relativity. To which Einstein commented *"Raffiniert ist der Herrgott…"* The experiment was wrong and the aphorism is in my New Yorker profile. In 1930 the mathematician Oswald Veblen who heard Einstein say this asked and got his permission to make this part of the fireplace in Fine Hall. The mathematics department has moved but the aphorism has remained where it was.

# A Little List                    19

Kanchenjunga from near Darjeeling

I guess summing up is a normal human instinct. We all make lists. At least I do. When I was living in Paris in the 1960's, I made a list of all the Michelin starred restaurants I had eaten in. Because of the very favorable currency exchange, by the end it was nearly all. I am now going to discuss two of my lists. The first involves the ten highest mountains in the world. At some point I decided that I wanted to see all of them. Climbing any of them was out of the question, but at least I could see them. I will give you the list and explain how I managed to see all of them. The second of my lists was not something that I set out to do. It consists of most of the founders of the quantum theory and how I saw them. I will explain.

The ten highest mountains in the world are in Asia and can be found, at least parts of them as they often constitute boundaries, in four countries, India, Nepal, Tibet and Pakistan. I will begin with the highest, Everest, which is on the boundary between Nepal and Tibet. Its altitude is now given as 29,035 feet. I have seen it from both countries and I will explain the circumstances. In 1967, I received a commission from the New Yorker magazine to travel to Nepal and to report on the country. I recruited a Chamonix mountain guide and his wife. We planned to trek in the high country, which in Nepal is very high and their experience would be helpful. At the time, large commercial airliners could not land in Nepal, so in September of 1967, we flew first to New Delhi and then in a Nepal Airlines plane practically due east to Kathmandu. I will never forget my first sight of the Himalayas. It was the end of the monsoon season and we first flew over flooded rice paddies on India. After some time, I noticed to the north some clouds that were at a greater altitude than the plane. But as I kept looking, these clouds did not move and I realized that they were mountains. The three of us had spent much of our lives in the mountains, but we had never seen anything like this. They were overwhelming. These were the western Himalayas, while Everest is north-east of Kathmandu and not visible from it. To see Everest, we trekked for days with a crew of Sherpas and porters. Our first goal was to get to Namche Bazaar, the Sherpa capital. The last step was to climb up a steep trail. At one point I saw a collection of prayer flags above me. This always means some awe inspiring view. In this case it was Everest! There it was in its gigantesque glory. I had to be pried away. In later years, I had the chance to fly over it and realized what an enormous land mass it is. In 1987, I visited it from the Tibetan side and even climbed up to about 20,000 feet. This was the side the early British expeditions tried to climb. In 1953, the successful British expedition climbed it from the Nepalese side and in 1967 we visited their old base camp.

K2 (Mount Godwin Austin) at 28,250, which is on the border of Pakistan and China, is the second highest mountain in the world. 'K' stands for 'Karakoram'. There is a K1 (Masherbrun) at 25,659 feet, which was for a while thought to be the highest mountain in the Karakoram range. Unlike Everest, which is now a tourist mecca, to get to K2 requires a small expedition. It is by the way, technically a much more difficult climb than Everest. My sighting of it was something of an accident. In 1969, I was awarded a Ford Foundation visiting professorship at what is now known as the Quaid-i-Azam university, in the Pakistani capital Islamabad. I decided that the obvious way to get there was to drive from France to Pakistan in a Land Rover. I once again recruited my guide friend and his wife. In the event, our route took us to Italy, Yugoslavia, Greece, Turkey, Iran, Afghanistan and over the Khyber Pass to Pakistan. Upon getting to Rawalpindi, I got the news that the university was closed for a month. This was very good news indeed, since we were then free to explore the Northwest frontier of Pakistan. In the event, we visited places like Swat, Chitral and Gilgit. On a few visits, we could drive in the Land Rover. But in most of them, we flew. This was true of our visit to Gilgit, which has a substantial community of over 200,000 inhabitants. It is also the gateway to many of the high mountains. On the flight from Islamabad to Gilgit, we found ourselves to be the only foreigners. At some point, the stewardess said that the pilots had invited us to the cockpit for a better view. When we got there, one of the pilots asked if we would like to see K2. We said that nothing would make us happier and he pointed the nose of the plane upward. At some point the most incredible black rock spire appeared — K2. It looked menacing and very dangerous. We stood there with our mouths open when the stewardess came in to announce that one of the passengers had just fainted. The plane did not have a very functional air pressure system. He took it to a lower altitude and K2 vanished.

Kanchenjunga at 28,169 is the third highest mountain in the world. It is on the border between India and Nepal in the extreme west of

Nepal. I cannot read its name without a certain feeling of regret. During the many years that I trekked in Nepal, I do not remember any trekking agency offering treks to Kanchenjunga. Perhaps it was too remote and the logistics too complicated. I now see that treks are offered and some seem to propose the possibility of stays in tea houses. The treks seem to last for a month and trekkers are warned that they are quite rigorous. But I do not regret not trekking to Kanchenjunga. As I will now explain, I regret not going to Darjeeling, the Indian hill station which is south of the mountain and from which you can find places to get stunning views.

I had read about Darjeeling since I was a boy. I knew the British were fond of it to get away from the heat in the plains. It is at nearly 7,000 feet. I knew that there was a funky train that got you up there and I knew that some of the most famous Sherpas including Tenzing, who with Hillary first climbed Everest, came from there. I always thought that if I got the chance I would visit it. But in 1988, I got the chance and didn't. This is my regret. Here is how it happened. In 1988, I took a trekking trip to Bhutan. It was a fascinating beautiful country but our trek was somewhat disorganized and by the end no one liked anyone else very much. There had been an offer of an after trip to Darjeeling but by this time I only wanted to get home. We drove down to India and on the way, there was a turn off that said Darjeeling up the hill. I felt a real pang of remorse. Here is how I did see Kanchenjunga in 1969.

After the fall term at the university in Islamabad, there was a vacation of a couple of weeks. I thought that I would use it to go back to Nepal. I had friends there and it would make a cheery Christmas. There was no way to fly there directly from Islamabad. Given everything the best way was to fly to Dacca, in what was then East Pakistan. Spend the night and then fly to Kathmandu which is to the north west. Not long after the plane took off, we saw an incredible mountain — Kanchenjunga — and flew alongside it for some time. It was a sight you don't soon forget.

The fourth highest is Lhotse at 27,940. It is really an appendage to Everest so we saw it constantly on our trek. The fifth highest is Makalu at 27,766. It is not far from Everest and there are various vantage points from which you can see it. The sixth highest is Cho Oyu at 26,906. It is again on the Nepal-Tibet border. If you drive from Kathmandu to Lhasa on the Friendship Highway, which I did twice, there is a period when the mountain is in full view from the road. You can't miss it. Likewise, if you trek in Western Nepal you can't miss Dhaulagiri at 26,795 and if you trek in the Annapurna region of western Nepal you can't miss Manaslu at 26,781. Naga Parbat is the western anchor of the Himalayas. It lies in Pakistan. It is 26,550 feet high. It stands in splendid isolation. Its rise in elevation from its base is as much as any mountain. I saw it from the air on our flight to Chitral, which is close to the border with India. It is a magnificent object and one can almost forget the number of climbers who have lost their lives on it.

The last mountain on the list, Annapurna at 26,545 is my sentimental favorite. It is the first 8000-meter mountain to be climbed to the summit. It was done so by the French in 1950. I got to know three members of the expedition, Maurice Herzog, Lionel Terray and Gaston Rébuffat. The latter two were Chamonix guides and Herzog was selected as the leader of the expedition. I did a climb with his brother Gérard. The two brothers were barely on speaking terms in good measure, because Gérard was not selected for the expedition. Terray and Rébuffat were not on speaking terms either. There was an incident high on the mountain, where they had to spend a night freezing in a crevasse. Terray tried to tend to everyone while Rébuffat was only interested in himself. Herzog came several times to Colorado and since I speak French I helped interpret. Terray was hard to know but Rébuffat was very friendly and I came to like him. Herzog's book on Annapurna is a classic but it does not reveal everything.

About seeing Annapurna there is a complication. Annapurna is really the name of a range. The French climbed Annapurna1 which is the highest and most visible. I first saw it in 1967, when after our Everest trek we did a shorter one in its foothills. We called a halt when it began to snow seriously. But some years later, I did the classic Annapurna circuit trek. This lasted over a month, during which we walked over a hundred miles. We saw the rest of the range. Some of the western side of the trek can now be shortened, by taking advantage of a motorable road. The trek is so popular, that there were even then tea houses and little hotels on the way. Nonetheless it was extraordinarily beautiful.

So those are my ten mountains. Now to the physicists.

The German physicist Max Planck introduced the notion of the quantum of energy just at the beginning of the $20^{th}$ century. He died in 1947, so I did not get to see him. In 1905, Einstein brought the quantum into physics. As my first great teacher in physics Philipp Frank put it, it was like beer. What Planck said, was like the proposition that whenever you buy beer, it was always sold in pints and quarts while Einstein said, that wherever you find beer, it is always in pints and quarts. I got a glimpse of Einstein once. I had a friend who had gone for a year to the Institute for Advanced Study in Princeton where Einstein was a professor. I had gone to visit him. Late in the afternoon, we were in front of Fine Hall where Einstein had his office, when he emerged and walked to the waiting Institute station wagon. I admit this is not much of a viewing, but I am somehow reminded of a joke that the physicist I. I. Rabi once told me about a lion hunter. When asked how many lions he had actually killed he replied, "None, but that is a great many when you are talking about lions." I can do better with Niels Bohr, who after Einstein was the greatest physicist of the $20^{th}$ century.

Bohr had a standing invitation from Robert Oppenheimer, the director of the Institute, to visit whenever he liked. He visited in the spring of 1958, when I was finishing the first of what turned out to be a two year stay at the Institute. Oppenheimer decided that we should have a little seminar, in which we should all tell Bohr what we had been doing. Normally I would have pleaded almost anything to get out of this, but as it happened, I had been working with two recent Nobelists, T. D. Lee and C. N. Yang. They did not want to talk about this work, but about their own work so they nominated me. I was so nervous, that I took about three minutes to give a totally incomprehensible talk. After it, Bohr said that it was "very interesting", which was Bohrian for saying that it wasn't. If it had been really interesting to him, he would have drilled me until both of us had a real understanding of the subject.

Bohr's work on the quantum theory was the bridge between classical physics and the real quantum theory, which began with Werner Heisenberg in 1925. Heisenberg did not understand what he had done at first but it opened up a new world. Heisenberg remained in Germany during the war and worked on nuclear energy. Whether he and his group had an atomic bomb in mind has been the subject of controversy ever since. He was not viewed with much affection by the people who had worked on our atomic bomb, many of whom were refugees from Europe. So his visit to Cambridge and M.I.T. in the fall of 1949 aroused mixed feelings. He seems to have been received correctly by Victor Weisskopf, the leading theorist at M.I.T., and one of the Los Alamos refugees, who even gave a party for him which many people boycotted. At this time, I was beginning my junior year at Harvard and already knew enough physics, so that I wanted to hear Heisenberg's lecture, which I did. I recall that I found it incomprehensible. Needless to say I had not been invited to the party.

Erwin Schrödinger, not long after Heisenberg published his work, invented what at first seemed like a different quantum theory. Heisenberg's version was known as "matrix mechanics" and Schrödinger's version was known as "wave mechanics." But Schrödinger showed that these were representations of the same theory and you could choose which version you wanted to use depending on the problem. Schrödinger was a hero of mine not only for his work on the quantum theory, but for a wonderful brief book called "What is Life?" which tried to account for the scientific basis of biological reproduction. This book inspired a whole generation of scientists to go into biology. As it happened, I spent part of the spring of 1960 in Vienna at the Boltzmann Institute. My host was the late Walter Thirring. Thirring informed me that the Schrödingers were living in Vienna and that they welcomed visits by young people. I leapt at the chance. A small group of us went to the Schrödingers. It turned out that they were living in one of those classic Viennese apartment buildings that looked as if it came out of The Third Man. I expected to hear that music and to see Orson Wells at any moment. The ancient elevator with its open windows would have been an ideal place. The Schrödinger apartment was overflowing with books in every possible language. His wife who must have put up with a lot — Schrödinger always had mistresses who kept coming down with children — served us Viennese coffee and pastries. I don't recall the conversation but just before we left, he looked at us with his extraordinary blue eyes amplified by his glasses and said, "There is one thing we have forgotten since the Greeks…**modesty**!" I have no idea what he meant. My great regret is that I did not ask his thoughts about the double helix structure of DNA which bolstered his views. He died in January of 1961.

Paul Dirac was a true original. He invented his own version of the quantum theory and was the first person to make a successful union of the quantum theory and the theory of relativity. The off springs were anti particles, the first one to have been discovered was the anti-electron — the "positron." He too had an open invitation to visit the Institute and we knew that he had come to Princeton in the fall of 1958 but no one had seen him. We had a regular weekly seminar and when one of them was in process in walked Dirac. He was wearing some kind of high boots and rough clothes. It turned out that he was spending a good deal of time in the woods behind the Institute clearing some kind of path with an axe. As it happened I had several occasions to observe him at close range. When his wife was away he often had dinner with us bachelors. I remember that one of our group asked him if he ever collaborated. "The really good ideas are had by only one person," was his response.

Dirac was noted for his economy of words. Here is an example that I witnessed. Oppenheimer had decided that we would not have phones in our offices, so as not to impede our thoughts. So there was a hall phone which when it rang disturbed everyone. One of the Institute professors, Abraham Pais happened to be in my office when the phone rang. It was for Dirac. We could hear him apparently saying no to something. He spotted Pais in my office and came in to ask him a question. He had been called by a reporter from the New York Times, who had heard that Dirac was going to give a talk in New York. He wanted a copy of the manuscript which Dirac did not want to be made public before his talk. He asked Pais for advice and Pais said to write on the manuscript "Do not publish in any form." Dirac listened and then just stood there. Pais and I went on with our conversation. Suddenly Dirac piped up and asked, "Isn't "in any form" redundant in that sentence?"

Wolfgang Pauli was not one of the inventors of the quantum theory but he was one of the foremost practitioners. He was an eccentric man who liked to tease. He would attach a nickname to someone and it would never be removed. He did it to me. At the time I was a postdoctoral at Harvard. I was going to New York to attend a physics meeting at which Pauli was to give a lecture. Then he was coming to Cambridge. Julian Schwinger, the foremost theorist at Harvard, took me aside. He said that he and his wife wanted to give a party for Pauli and would I be so kind to find him at the meeting and deliver the invitation. This I did but apparently in a soft voice. Pauli began calling me the "whisperer" and never stopped.

I have left out of this list Max Born, because I never met him. Heisenberg was his assistant when he discovered matrix mechanics. Heisenberg did not at the time know what a matrix was until Born explained it to him. Born also introduced the probability interpretation of the quantum theory. He finally won the Nobel Prize in 1954 and used the money to reestablish himself in Germany, from which he had been expelled. He sounds like a very interesting man and I wish I had met him.

These are two of my lists and I am sure that I will come up with others.

# Anti-Semitism at Harvard    20

"All that lattice and no tomatoes…" Tom Lehrer

Donald Menzel, Einstein, George Birkhoff and Harlow Shapley's son Carl on the occasion of Einstein's receiving an honorary degree from Harvard in 1935.*

---

* Photo credit: Emilio Segre Archives — The American Physical Society

An anti-Semite is someone who dislikes Jews more than is absolutely necessary.

I received my master's degree in mathematics at Harvard in the spring of 1953. I do not recall that there was any kind of ceremony. I think it just came in the mail. That fall, I was summoned to the chairman of the department's office Garett Birkhoff and told I had to make a choice. He said that I was taking too many physics courses and if I continued to do that I should leave the department. I then left his office and went to the physics department and signed up for something I have never regretted. There were a least two reasons why I made the switch. In the first place, the kind of mathematics I found interesting — Hilbert space, tensor analysis, some group theory — always had a connection to physics. But more importantly, I did not think that I would become a very good mathematician. This is quite different from being a good math student, which I was. I had graduated *magna cum laude*, which was the reason I was accepted as a graduate student in the department. And I had even gotten an A in Birkhoff's algebra course, whilst I had also gotten to know some real mathematicians among my fellow students. And of course among the faculty. I knew I was not one of them. I felt that with guidance, I could get my PhD and that I might end up teaching math somewhere but that would be about it. So I decided that I would take my chances with physics.

There was something unlikable about Birkhoff.

I always referred to him as the permanently rotating chairman of the Harvard mathematics department. His great specialty was lattice theory and hence Tom Lehrer's song. I did not think he was a very good teacher and maybe I conveyed that. I think he was glad to get rid of me in the department. What I did not know about was his father.

George David Birkhoff was born in 1884 in Overisel, Michigan in 1884. The town was named after the Dutch province of Overijssel. His father David who was of Dutch origin was a medical doctor. I do not know whether the young Birkhoff acquired his anti-Semitism from his family, or at the University of Chicago where he studied briefly and then later took his PhD, or at Harvard where he took both his BA and MA. In 1912 he became a professor at Harvard, remaining for the rest of his life. In 1936, he became Dean of the Faculty of Arts and Sciences, which meant he had influence on faculty appointments. He made it his mission to see that no Jew would get a professorship in the Harvard math department. He even went as far as to oppose the candidacy of a Russian born Jew Solomon Lefschetz for the presidency of the American Mathematical Society. He wrote to a colleague in 1934,

"I have a feeling that Lefschetz will be likely to be less pleasant than he had been, in that from now on he will try to work strongly and positively for his own race, [Jews] are exceedingly confident of their own power and influence in the good old USA… He will be very cocky, very racial and use the Annals of Mathematics as a good deal of racial perquisite. The racial interests will get deeper as Einstein's and all of them do.[1]"

This was written in 1934 and the picture of Einstein and Birkhoff at Harvard was taken in 1935. I do not know if Einstein was aware of Birkhoff's anti-Semitism in 1935. At some point he is quoted as saying that "G. D. Birkhoff is one of the world's great anti-Semites." Did he feel this at the time the picture was taken? Does his body language tell us something?

**Endnote**

1. I thank Gino Segre for confirming this.

Twenty One **21**

Ed Thorp[1]

"Markets can remain irrational longer that you can remain solvent"
John Maynard Keynes

My first encounter with casino blackjack took place in the late summer of 1957. I was a summer intern at Los Alamos and I learned that it might be possible to go to Nevada and watch some nuclear bomb tests. I was not working on anything connected with nuclear weapons, but I wanted to be educated. Although I had a

---

[1] Photo credit: Mark Jordan

Q Clearance, which would have entitled me to know I had no "need to know" so no one told me anything. I had a senior colleague named Francis Low, who was in the same position and who had given me the idea of trying to go to Nevada. Francis, it turned out had done some preparation. This had nothing to do with nuclear weapons but rather with casino blackjack. He reasoned that we would be spending some time in Las Vegas and would visit casinos where blackjack was offered, so he should educate himself. He had come across an article entitled "The Optimum Strategy In Blackjack" which had been published in the September 1956 issue of Journal of the American Statistical Association. It had four authors who turned out to be soldiers stationed at the Aberdeen Proving Ground in Maryland. They were all mathematicians in various stages of their careers and this was something that they had done in their spare time. It turned out that none of them had actually played casino blackjack but they knew the rules and worked out how to apply them using the calculators on the base.

The rules for casino blackjack are nearly the same in all casinos — at least those in the United States. There is a dealer who might as well be an automaton. In Nevada I have encountered almost equal numbers of male and female dealers. There is a table around which usually seven players are seated facing the dealer. First bets are placed. There is usually a casino minimum.

If the deck of cards is fresh the dealer shuffles it and a player cuts the deck — exchanges top and bottom halves. The dealer then puts a card face up at the bottom of the deck generally not revealing what it is. Then each player is dealt two cards face down while the dealer takes two one of which must be face up. The fact that all the cards in the illustration are face up means that the hands have been played. Each card has a numerical value except the ace which has two (one or eleven) and the player can make the choice. Face cards count for ten. The idea is to get as close to twenty-one as possible. If you do it on the deal, this is called "blackjack" and you get paid a bonus, unless the dealer also has blackjack in which case, no money changes hands. When it is your turn,

you can decide to take another card which is dealt face up. If you go over twenty-one, the house wins and you are required to show your cards and pay the bet. It is in this that the dealer's advantage lies. The dealer plays last and although he or she may eventually go bust, the players who have preceded and have gone bust have already paid off their bets! You can, if you have not gone over twenty-one, take additional cards until you decide to stop. There are various nuances involving increasing your bet which I will not go into. When all the hands are played it is the dealer's turn. If the dealer had sixteen or less, he or she must take another card. If the dealer has seventeen or more, he or she must stop. The whole thing could be done by robots and can now be played on a computer. The dealer's final number is compared to those of the players and bets are paid off one way or another. There are nuances about ties which again I will not go into.

What the four soldiers did was to analyze all the combinations involving what card the dealer was showing and what cards the player was holding to determine what the player should do to optimize each choice. There is no guarantee that you will always win but you will lose, they claimed, as little as possible. Francis made an estimate which went something like this, Suppose I play for an hour (say twenty games) a new game every three minutes and suppose I bet ten dollars a game. Then using their optimum strategy, if my advantage is about one percent, I would make about two dollars an hour but at least one would have beaten the casino. At the end of the paper there is a caveat — which turns out to be big enough to drive a truck through. They write,

"The optimum strategy was developed under the assumption that the player does not have the time or inclination to ultilize the information available in the hands of players preceding him in the draw. This information is non-existent when the player sits on the dealer's left and is greatest when the player sits on the dealer's right. [One then sees which cards have been played.] There are tremendous difficulties, however, in using this information except in an intuitive and non-scientific manner."

I will come back to this when I briefly wrap up my Los Alamos experience. The people at Los Alamos were well-aware of this article and they ran millions of hands on a mainframe computer to check the results. Having satisfied themselves that they were correct, they were summarized on a little folder we were all given. I had a chance to use it almost from the time the plane landed in Las Vegas. I found that the casino game went so fast that the folder was not much of a help. Later I managed to contact one of the authors of the paper and he was totally unaware of the Los Alamos effort. But the paper had a certain readership among mathematicians, including Robert Sorgenfrey who was a professor at UCLA. He had a younger colleague named Edward O. Thorp, who was planning to take a Christmas vacation with his wife in Nevada. Sorgenfrey gave Thorp the article and he decided to try it out with ten silver dollars. After he lost them, he decided there must be a better was to play the game and what he found changed the nature of blackjack forever.

Thorp rejected the idea expressed in the paper "that the player does not have the time or inclination to utilize the information available in the hands of players preceding him in the draw," and realized that if this information was used correctly the player could actually win. The general rubric under which this goes is "card counting" (counting the cards that have been played before it is one's turn). Because of movies like '21', card counting in blackjack rouses the image of packs of MIT students with eidetic memories, descending on Las Vegas like plagues of locusts. There is card counting and there is card counting. Card counting 101 can be practiced by anyone who knows how to count. It consists simply of counting the fives that have been played. Why fives? From the point of view of the dealer fives are extremely valuable. Remember that if a dealer holds cards that add up to sixteen or less, he or she must take cards until at least seventeen is reached and then must stop. This means that if the dealer holds sixteen or less and takes a five there is no way the dealer can go "bust", to use the term of art, and there are several ways to draw to high numbers. Thus the knowledge that there are no more fives for the dealer to draw is very valuable, and that knowledge can sometimes be determined by observing the cards that have been played. In general, one can get an idea of the status of the fives. Shortly after having this insight, Thorp moved to MIT. He taught himself FORTRAN, so he could program the then new MIT mainframe IBM 704 computer to carry out the millions of calculations he needed to test his method. Prior to this he had contacted Roger Baldwin the soldier who was in charge of the first effort. Baldwin sent him the results of their calculations which they had done on manual calculators. In 1960, Thorp presented a paper at the American Mathematical Society meeting in Washington D.C., "Fortune's Formula: A Winning Strategy in Blackjack", which combined the work of the four soldiers with his new insight involving card counting. The paper created a sensation in what one would imagine must have been an otherwise rather staid meeting. Thorp was asked to give interviews. It was written up by no less a writer than Tom Wolfe. He received a variety of offers from people who wanted to back him on a trial run. He made a selection whom he later identified only as "Mr. X" and "Mr. Y", who put up a $10,000 stake. It turned out that "Mr. X" was a professional gambler with mob associations named "Manny" Himmel. "Mr. Y" was Eddie Hand a wealthy business man from upstate New York. Thorp and Himmel headed for Las Vegas for the trial run, later to be joined by Hand. They wanted to put up a hundred thousand dollars, but this seemed extravagant considering the risks, so ten thousand was agreed to. By this time Thorp had improved his system so he was counting both fives and tens. When Thorp began winning, they were politely but firmly told to leave many of the casinos they were in. Sometimes they were even offered a free dinner. They ended up with winnings of twenty-one thousand dollars. Manny, who could not resist playing, and did not use the system lost a good deal of his own money. Thorp wrote some of this up in a book called Beat the Dealer, which was an immense best seller and is still in print in new editions. Casinos have now brought to bear defenses, such as having multiple decks and frequent shuffling which makes card counting more difficult. Thorp also discovered that some of the casinos cheat in blackjack. If a dealer is a very good card handler, he or she can determine the top card they are about to deal and the one beneath, and if that one is less favorable to the player to deal that one. Some of Beat the Dealer deals with casino cheating.

In Beat the Dealer, and in some of Thorp's other books, there are autobiographical snippets but now he has written a full-scale autobiography, A Man for All Markets. It is a fascinating book and one comes away marveling at the odd twists of fate that have governed his life. He was born on August 14, 1932 in Chicago. His father had fought and had been wounded in France in the First World War. He then attended Oklahoma A&M for a year and a half, before being forced to drop out because of lack of money. But it left him with a lifetime intellectual curiosity which he passed on to Thorp. He got a job as a security guard at a bank with a salary of 25$ a week. Prior to that he had had a job in the constabulary in Corregidor in the Philippines. There he met Thorp's mother. Her father had left Germany to become an accountant for the

Rockefellers. Fortunately, the young couple came to the United States before the Japanese invasion of the Philippines. The rest of the family were interred. It became clear that young Thorp was exceptionally bright but this was not evident at first, because until the age of three he refused to talk. He was examined by a physician and when asked to point to various objects, he did so without difficulty. Thorp recalls being taken to a department store in Chicago and a woman friend of his mother, pointing to an elevator where a man and two women had just gotten off, asked where they were going to which Thorp replied, "The man is going to buy something and the two women are going to the bathroom to do pee-pee." Not only could he talk but he could count. In fact, Thorp had a love affair with numbers. He loved big numbers and could do elaborate mental arithmetic. He soon learned to read and his father supplied him with all the books he could handle. After Pearl Harbor, the family moved from Chicago to Southern California with the hope of finding war work. Indeed, during the war his mother worked as a riveter.

Despite the fact that both his parents were working, the family was not well-off. There was no question of Thorp going to any private school. It was clear that he was way ahead of his age group in school. To find a suitable placement, he was asked to take what turned out to be an IQ test. He scored the highest ever recorded in that school and began high school a couple of years early. Fortunately, he attracted the attention of a teacher who was able to guide him in more advanced work. He also began selling newspapers and putting the money he earned into war bonds to be used later for college. He managed to get a ham radio license, which meant that he learned to transmit and receive Morse code at a rate of twenty-one words a minute. But his real interest was chemistry. He loved explosives. In the meanwhile, the surviving members of his mother's family arrived from the Philippines to live with them, putting an even greater strain on the family resources. It became clear that without some outside help, there was no way that Thorp could go to college. At this time, the Southern California Chemical Society gave an annual examination which gave the winner a scholarship for college. Thorp decided to take it as an underage junior in high school. He could not afford a decent slide rule and the last part of the exam required one to do the problems. Using mental arithmetic, he came in fourth in the city, which not only did not qualify him for a scholarship, but he was not allowed to repeat the exam the next year. He decided to take the one in physics, and now armed with a proper slide rule, he won. This meant that his college tuition would be paid and now he had to decide where to go. Given his abilities and interests, a natural place would have been Cal Tech in Pasadena. But the living costs were too high, so he chose to go to Berkeley where he spent a miserable year.

By this time, his mother and father had divorced. It turned out that she had been having an affair. She had also sold his war bonds and had pocketed the money. His father could only afford forty dollars a month to help out. Thorp worked and shared quarters with multiple roommates. In addition, he did not think that the classroom lectures were very good, so he switched to UCLA. He got his PhD in mathematics at UCLA and then got an offer of an instructorship at MIT. In the interim he had gotten married.

During his stay at MIT, Thorp encountered Claude Shannon, who was a professor there. Shannon was one of the true pioneers of the information age, but he had a somewhat mischievous side. He tried to find the smallest unicycle one could ride and he invented a juggling machine, and learned to juggle himself eventually on a unicycle. He became interested in roulette. He realized that the balance of a roulette wheel could never be perfect, so that if you could study it in operation, you could learn the slight tendencies that the ball, that circulates opposite to the rotation of the wheel would have, to land on certain numbers. This was just the sort of thing that was up Thorp's alley. In fact he had studied roulette, as well as blackjack and backgammon. The two of them invented a wearable device that could measure these deviations, which they tested on a roulette wheel in Shannon's basement. They then went to Nevada and saw that it worked there, although it kept breaking. It was by the way, Shannon who had gotten Thorp's paper on card counting published in the *Proceedings of the National Academy of Sciences*, not a very likely subject for the Academy journal.

In 1961, Thorp was offered a third year at MIT, but he felt that the prospects of eventually getting tenure there were very uncertain. He got an offer from the New Mexico State University in Las Cruces, which was in the process of building a PhD program in mathematics. It carried tenure and a low teaching load, so he accepted and spent four very happy years there. But then he got an offer from a new campus of the University of California at Irvine, where he spent the rest of his academic career. If all you were to look at was the list of Thorp's papers in pure mathematics, such as 'Projections onto the subspace of compact operators", you would be persuaded that this was the academic output of a pure mathematician. You would not get a hint that you were dealing with one of most successful investment counselors who ever lived.

In the summer of 1968, Thorp first met Warren Buffett. (They are almost the same age, Buffett being two years older.) It was at the house of Ralph Waldo Gerard, who was the dean of the graduate school at Irvine. The Gerard's had been early investors in a Buffett partnership, one which he was planning to dissolve. Thorp was being vetted as a possible replacement. Buffett gave his favorite example of compound interest. If the Indians, who sold Manhattan for the 24$ worth of trinkets, had been able to invest the 24$ in something that compounded at 8%, they could now buy Manhattan back with all the improvements. Buffett decided that Thorp would do, despite the fact that their investment strategies were very different. Buffett looked for undervalued companies, which he could buy a large or even controlling interest in. If he liked the management, he would wait patiently for the company to grow. But in 1968, he was having difficulty in finding such companies, so he was dissolving this partnership. Thorp, on the other hand, was interested in the use of "derivatives". Derivatives are not actual stock, but might be the options to buy or sell a stock at some future date. There were issues involved that appealed to Thorp as a mathematician. What were options actually worth, and could you hedge them against loss? Let me give a very simple example.

Suppose you know about a stock, which say in one month will either have a value price of $120 or $80. Its current value price is $100. Further suppose that you know that the probability of the higher price is ¾, so that the probability of the lower price is ¼. You want to buy an option to purchase the stock in say one month at its current value price of $100. You can always buy the stock at its future value price. This does not require an option. What is such an option worth? If the stock price becomes $80, you will not exercise your option, but if it becomes $120 you will. Hence, your expectation for the option value is 3/4x$20+0x$80=$15, and this seems like a reasonable price to pay. If you pay $15 for the option, you have a high probability of earning $5. In reality, the probability situation is never clear cut, and the heavy lifting for the analyst is to estimate it. But you can do much better as this example will illustrate, if you allow "hedging", employ "arbitrage" how to replicate the option in a more sophisticated way.

Let us suppose that there is a friendly bank that is willing to lend me money interest free. With more work you can include the effect of interest. This activity is called "arbitrage." For the sake of this example, I will suppose stocks can be purchased in half fractional shares. The option pays $20 if the stock moves up and $0 if the stock moves down. A half share of stock pays $60 if the stock moves up and $40 if the stock moves down. If you buy a half share of the stock which costs $50 with 10$ of your own and 40$ of borrowed money you will be worth $60 and owe $40, That is exactly $20 if the stock moves up. Similarly, you will be worth $40 and owe $40 if the stock moves down. That is exactly 0$. These final values are identical to those of the option. Thus you can replicate the payoff of the option exactly, by borrowing $40, investing $10 of your own, and buying the stock. Therefore, it costs you only $10 of your own money to manufacture the option and that should be its fair price, lower than the $15 of the expected value. I can take the $15 from you to buy the option, but I will put $5 in my pocket. By borrowing from the bank and buying the half share I can replicate the option and make $5. You will never know the difference. This sort of replication exponentiated is what hedge fund managers and options traders do on a daily basis.

To evaluate options, you need some way of determining the probable future value distribution of stock prices. The first person to attempt to do this with some rigor was a French mathematician named Louis Bachelier, who in 1900 published a thesis which he called *Théorie de la Spéculation*. Bachelier was born in 1870 in le Havre. The reason that it had taken so long to do his thesis, was that his father a wine merchant had died and Bachelier had family responsibilities. It seems that he had worked in the interim on the Paris bourse. He had also heard some lectures on probability by the great French mathematician Henri Poincaré, who became one of the referees of Bachelier's thesis. Bachelier's starting principle was that at any stage, it was equally likely for a given stock to go up as down. Someone unfamiliar with random walks might be inclined to ask, how then can the stock price go anywhere? But after the first move, the stock is again equally likely to increase, as it is to return to its previous value. In fact, it turns out that it departs from its initial value by an amount that is proportional to the square root of the time elapsed. I have not been able to learn if Bachelier made any investments based on his work, or indeed if anyone else did at the time. Remarkably the next person to make use of these general ideas was Albert Einstein.

There is no evidence that Einstein ever heard of Bachelier. He certainly did not have any interest in stock speculation. What he did have interest in, is what is known as "Brownian motion." In 1827 the Scottish botanist Robert Brown noted that microscopic particles, for example pollen grains, in liquids engaged in an odd motion. He reasonably concluded that these grains were alive. But many other kinds of microscopic objects were examined and they too exhibited this motion although they could not possibly be alive. Einstein decided that these particles were being set into random motions by the collisions they had with the atoms of the liquid. When experiments confirmed this 1905 speculation, much of the objection to the existence of atoms was put to rest. During the next years, mathematicians either re-discovered or built upon Bachelier's work. In 1950 a noted mathematical statistician named I. J. Savage sent out post cards to some of his colleagues describing his discovery of Bachelier. One of the people that he sent cards to was Paul Samuelson of MIT. Samuelson was the first American to win the Nobel Prize in economics. He was deeply impressed by what Bachelier had done and communicated it to colleagues. Among them was Paul Cootner, who put together an anthology called The Random Character of Stock Market Prices, which contained an English translation of Bachelier's thesis. This was published in 1964. Thorp read it in 1967 and proceeded to produce a practical formula for estimating option prices. He was not interested in academic economics, but how to use it to make money. So he never published the formula keeping his techniques proprietary. A few years later Fischer Black and Myron Scholes by one method, and Robert Merton by another, rediscovered the method and made it mathematically rigorous. Scholes and Merton shared the Nobel Prize; Black had died.

Thorp remained a professor in the mathematics department until 1977. Then he became associated with the School of Management and taught courses in the mathematics of investment until 1982. At first, he invested for himself and for a few people he knew. But it became a full-time job. In 1974, he created something called the Princeton/Newport partners. There was one branch on the east coast and one in California. Fortunately for Thorp, these branches had a certain amount of independence. This was fortunate because on November 6,1989 a Federal District Judge in Manhattan sentenced the Princeton managing partner and several associates to jail, for several months for creating illegal tax losses. Thorp knew nothing about these, but he liquidated the fund and re-established himself as Edward O. Thorp and Associates. He writes about this candidly in the book. I have read that his net worth is about 800 million dollars.

In 2003, Thorp and his wife Vivian donated a million dollars to found a chair of mathematics at Irvine. The donation was in the form of shares from Warren Buffett's Berkshire Hathaway partnership. The idea was that these shares would appreciate so the available amount would be sustained. They stipulated that none of the recipient's salary should come from the donation. They felt that if the occupant was not good enough to be paid by the university, he or she should not have been hired in the first place. The present occupant is Karl Rubin, whose mother the great astronomer Vera Rubin just died. Thorp's wife Vivian died of cancer in 2011.

I met Thorp only once. At that time, and for reasons I can no longer explain, I was making yet another attempt to understand the mathematical philosophy of Immanuel Kant. I had started with the question of how much mathematics did Kant actually know. I concluded that as a young man he had studied and taught mathematics and had learned enough so that he could understand at least some of the physics of Newton. But I also came to realize that there was a mathematical problem, which seemed to be well known and which was under his very feet, for which he never seemed to show any interest. Kant spent his entire life within close proximity of the then Prussian city of Königsberg, his birthplace. He was a man of very regular habits, and each day after his noon meal, he went for a brisk walk. Below is a map of the seven bridges of Königsberg.

In his walks, Kant must have crossed these bridges many, many times. But as far as I can tell it never occurred to him to ask whether he could find a path that crossed all seven bridges once and only once. It did occur to many other people including the mayor of Danzig (the capital of Prussia), Carl Ehler, who was a mathematician. Ehler was stuck, so he appealed to one of the greatest mathematicians who ever lived, Leonard Euler. Euler was Swiss by birth, but at that time was working in St. Petersburg. I do not know whether he had ever visited Königsberg. Euler responded with a crisp message,

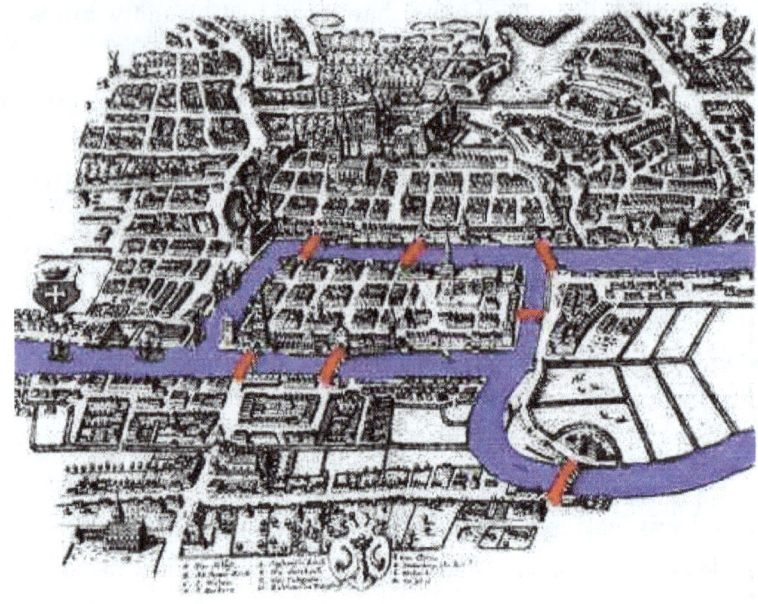

"...Thus you see, most noble Sir, how this type of solution bears little relationship to mathematics, and I do not understand why you expect a mathematician to produce it, rather than anyone else, for the solution is based on reason alone, and its discovery does not depend on any mathematical principle. Because of this, I do not know why even questions which bear so little relationship to mathematics are solved more quickly by mathematicians than by others."

Nonetheless he solved it and in the course of finding the solution created a new branch of mathematics: the theory of planar graphs.

He began by abstracting the situation down to its essentials.

He was then able to characterize any crossing from one landmass to another by a sequence of letters. He then showed that no sequence of letters can lead to a path that crosses all seven bridges only once. The interested reader can find the details on the web. Kant was apparently influenced by some of Euler's writings on physics. He seems to have tried to initiate a correspondence. There is no indication that Euler

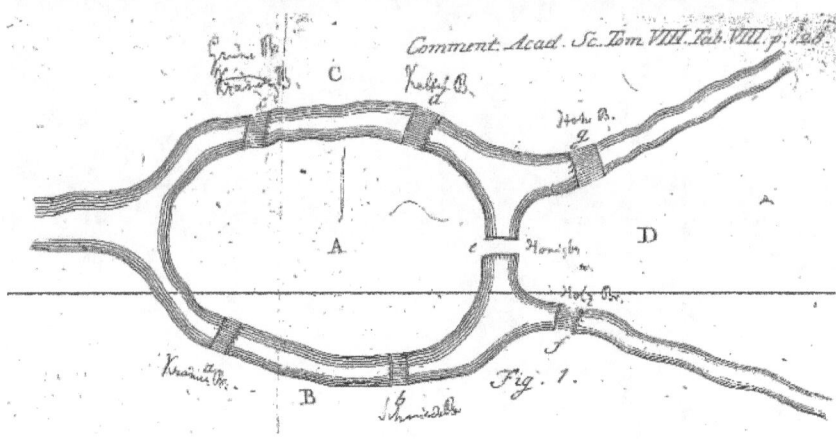

showed any interest. He certainly did not communicate to Kant his solution to the problem of the seven bridges of Königsberg.

In any event when I met Thorp I had some version of these diagrams. I asked him if he had ever heard of the problem. He said that he had not and then lapsed into what appeared to me to be a sort of meditative trance. After not many minutes he came up with a solution. I was once again reminded of the fact that while anyone can learn to do some mathematics, to really create it requires a special gift.

www.ingramcontent.com/pod-product-compliance
Lightning Source LLC
Chambersburg PA
CBHW061941220426
43662CB00012B/1983